対訳
ISO 9004:2018
(JIS Q 9004:2018)

ポケット版

品質マネジメント―組織の品質
持続的成功を達成するための指針

日本規格協会　編

*著作権について

本書は，ISO 中央事務局と当会との翻訳出版契約に基づいて刊行したものです．

本書に収録した ISO 及び JIS は，著作権により保護されています．本書の一部又は全部について，当会及び ISO の許可なく複写・複製することを禁じます．ISO の著作権は，下に示すとおりです．

本書の著作権に関するお問合せは，当会販売サービスチーム (Tel. 03-4231-8550) にて承ります．

© ISO 2018

All rights reserved. Unless otherwise specified, or required in the context of its implementation, no part of this publication may be reproduced or utilized otherwise in any form or by any means, electronic or mechanical, including photocopying, or posting on the internet or an intranet, without prior written permission. Permission can be requested from either ISO at the address below or ISO's member body in the country of the requester.

ISO copyright office
CP 401 • Ch. de Blandonnet 8
CH-1214 Vernier, Geneva
Phone: +41 22 749 01 11
Fax: +41 22 749 09 47
Email: copyright@iso.org
Website: www.iso.org
Published in Switzerland

本書について

本書は,国際標準化機構(ISO)が 2018 年 4 月に第 4 版として発行した国際規格 ISO 9004:2018 (Quality management − Quality of an organization − Guidance to achieve sustained success),及びその翻訳規格として,日本工業標準調査会(JISC)の審議を経て 2018 年 12 月 20 日に経済産業大臣が制定した日本工業規格 JIS Q 9004:2018(品質マネジメント−組織の品質−持続的成功を達成するための指針)を,英和対訳で収録したものです.

収録に際して JIS の解説は省略しています.JIS の解説を参照したい場合は,JIS 規格票をご利用ください.

規格をより深く理解し実践したい方には,日本規格協会より発行予定の解説書籍を併読されることをお勧めします.

また,本書内で引用されている ISO 9000:2015(JIS Q 9000:2015)については,姉妹書籍『対訳 ISO 9001:2015(JIS Q 9001:2015)品質マネジメントの国際規格[ポケット版]』(日本規格協会)を参照されることをお勧めします.

2019 年 2 月

日本規格協会

Contents

ISO 9004:2018
Quality management — Quality of an organization — Guidance to achieve sustained success

Foreword ··· 10
Introduction ···································· 16
1 Scope ·· 22
2 Normative references ················ 22
3 Terms and definitions ················ 24
4 Quality of an organization and sustained success ····················· 26
4.1 Quality of an organization ············ 26
4.2 Managing for the sustained success of an organization ························· 32
5 Context of an organization ········ 36
5.1 General ······································ 36
5.2 Relevant interested parties ·········· 38
5.3 External and internal issues ········ 40
6 Identity of an organization ········ 44
6.1 General ······································ 44
6.2 Mission, vision, values and culture ······· 44

目　次

JIS Q 9004 : 2018
品質マネジメント−組織の品質−
持続的成功を達成するための指針

まえがき ……………………………………… 11
序文 …………………………………………… 17
1 適用範囲 ………………………………… 23
2 引用規格 ………………………………… 23
3 用語及び定義 …………………………… 25
4 組織の品質及び持続的成功 …………… 27

4.1 組織の品質 …………………………… 27
4.2 組織の持続的成功のためのマネジメント … 33

5 組織の状況 ……………………………… 37
5.1 一般 …………………………………… 37
5.2 密接に関連する利害関係者 ………… 39
5.3 外部及び内部の課題 ………………… 41
6 組織のアイデンティティ ……………… 45
6.1 一般 …………………………………… 45
6.2 使命，ビジョン，価値観及び文化 …… 45

7 Leadership ... 48
- 7.1 General ... 48
- 7.2 Policy and strategy ... 52
- 7.3 Objectives ... 58
- 7.4 Communication ... 60

8 Process management ... 62
- 8.1 General ... 62
- 8.2 Determination of processes ... 66
- 8.3 Responsibility and authority for processes ... 70
- 8.4 Managing processes ... 72

9 Resource management ... 82
- 9.1 General ... 82
- 9.2 People ... 86
- 9.3 Organizational knowledge ... 94
- 9.4 Technology ... 96
- 9.5 Infrastructure and work environment ... 98
- 9.6 Externally provided resources ... 104
- 9.7 Natural resources ... 106

10 Analysis and evaluation of an organization's performance ... 110
- 10.1 General ... 110
- 10.2 Performance indicators ... 112
- 10.3 Performance analysis ... 118

7 リーダーシップ …… 49
7.1 一般 …… 49
7.2 方針及び戦略 …… 53
7.3 目標 …… 59
7.4 コミュニケーション …… 61

8 プロセスのマネジメント …… 63
8.1 一般 …… 63
8.2 プロセスの決定 …… 67
8.3 プロセスの責任及び権限 …… 71
8.4 プロセスのマネジメント …… 73

9 資源のマネジメント …… 83
9.1 一般 …… 83
9.2 人々 …… 87
9.3 組織の知識 …… 95
9.4 技術 …… 97
9.5 インフラストラクチャ及び作業環境 …… 99
9.6 外部から提供される資源 …… 105
9.7 天然資源 …… 107

10 組織のパフォーマンスの分析及び評価 …… 111
10.1 一般 …… 111
10.2 パフォーマンス指標 …… 113
10.3 パフォーマンス分析 …… 119

10.4	Performance evaluation	122
10.5	Internal audit	128
10.6	Self-assessment	134
10.7	Reviews	136
11	**Improvement, learning and innovation**	138
11.1	General	140
11.2	Improvement	140
11.3	Learning	144
11.4	Innovation	150

Annex A (informative) **Self-assessment tool** ································· 158

Bibliography ································· 296

10.4	パフォーマンス評価	123
10.5	内部監査	129
10.6	自己評価	135
10.7	レビュー	137

11 改善,学習及び革新 ……………………………… 139

11.1	一般	141
11.2	改善	141
11.3	学習	145
11.4	革新	151

附属書A（参考）自己評価ツール …………… 159

参考文献 ………………………………………………… 297

Foreword

ISO (the International Organization for Standardization) is a worldwide federation of national standards bodies (ISO member bodies). The work of preparing International Standards is normally carried out through ISO technical committees. Each member body interested in a subject for which a technical committee has been established has the right to be represented on that committee. International organizations, governmental and non-governmental, in liaison with ISO, also take part in the work. ISO collaborates closely with the International Electrotechnical Commission (IEC) on all matters of electrotechnical standardization.

The procedures used to develop this document and those intended for its further maintenance are described in the ISO/IEC Directives, Part 1. In particular the different approval criteria needed for the different types of ISO documents should be noted. This document was drafted in accordance

まえがき

(ISO の Foreword と JIS のまえがきは，それぞれの原文において内容が異なっているため，対訳となっていないことにご注意ください．)

　この規格は，工業標準化法第 14 条によって準用する第 12 条第 1 項の規定に基づき，一般財団法人日本規格協会（JSA）から，工業標準原案を具して日本工業規格を改正すべきとの申出があり，日本工業標準調査会の審議を経て，経済産業大臣が改正した日本工業規格である．これによって，**JIS Q 9004:2010** は改正され，この規格に置き換えられた．

　この規格は，著作権法で保護対象となっている著作物である．

　この規格の一部が，特許権，出願公開後の特許出願又は実用新案権に抵触する可能性があることに注意を喚起する．経済産業大臣及び日本工業標準調査会は，このような特許権，出願公開後の特許出願及び実用新案権に関わる確認について，責任はもたない．

with the editorial rules of the ISO/IEC Directives, Part 2 (see **www.iso.org/directives**).

Attention is drawn to the possibility that some of the elements of this document may be the subject of patent rights. ISO shall not be held responsible for identifying any or all such patent rights. Details of any patent rights identified during the development of the document will be in the Introduction and/or on the ISO list of patent declarations received (see **www.iso.org/patents**).

Any trade name used in this document is information given for the convenience of users and does not constitute an endorsement.

For an explanation on the voluntary nature of standards, the meaning of ISO specific terms and expressions related to conformity assessment, as well as information about ISO's adherence to the World Trade Organization (WTO) principles in the Technical Barriers to Trade (TBT) see the following URL: **www.iso.org/iso/foreword.html**.

This document was prepared by Technical Committee ISO/TC 176, *Quality management and quality assurance*, Subcommittee 2, *Quality systems*.

This fourth edition cancels and replaces the third edition (ISO 9004:2009), which has been technically revised. The main changes compared to the previous edition are as follows:
— alignment with the concepts and terminology of ISO 9000:2015 and ISO 9001:2015;
— focus on the concept of "quality of an organization";
— focus of the concept of "identity of an organization".

Introduction

This document provides guidance for organizations to achieve sustained success in a complex, demanding and ever-changing environment, with reference to the quality management principles described in ISO 9000:2015. Where they are applied collectively, quality management principles can provide a unifying basis for an organization's values and strategies.

While ISO 9001:2015 focuses on providing confidence in an organization's products and services, this document focuses on providing confidence in the organization's ability to achieve sustained success.

Top management's focus on the organization's

序文

(ISO の Introduction と JIS の序文は，それぞれの原文において内容が異なっているため，対訳となっていないことにご注意ください．)

この規格は，2018年に第4版として発行された **ISO 9004** を基に，技術的内容及び構成を変更することなく作成した日本工業規格である．

なお，この規格で点線の下線を施してある参考事項は，対応国際規格にはない事項である．

この規格は，**JIS Q 9000**:2015 で記載されている品質マネジメントの原則を参照しながら，組織が，複雑で，過酷な，刻々と変化する環境の中で，持続的成功を達成するための手引を提供している．品質マネジメントの原則は，一括して適用した場合，組織の価値観及び戦略のための統一的な基礎を提供することができる．

JIS Q 9001:2015 は，組織の製品及びサービスについての信頼を与えることに重点を置いているが，この規格は，組織の持続的成功を達成する能力についての信頼を与えることに重点を置いている．

トップマネジメントが顧客及びその他の密接に関

ability to meet the needs and expectations of customers and other relevant interested parties provides confidence in achieving sustained success. This document addresses the systematic improvement of the organization's overall performance. It includes the planning, implementation, analysis, evaluation and improvement of an effective and efficient management system.

Factors affecting an organization's success continually emerge, evolve, increase or diminish over the years, and adapting to these changes is important for sustained success. Examples include social responsibility, environmental and cultural factors, in addition to those that might have been previously considered, such as efficiency, quality and agility; taken together, these factors are part of the organization's context.

The ability to achieve sustained success is enhanced by managers at all levels learning about and understanding the organization's evolving context. Improvement and innovation also support sustained success.

連する利害関係者のニーズ及び期待を満たすための組織の能力に重点を置くことが，持続的成功を達成することについての信頼を与える．この規格は，組織の全体的なパフォーマンスへの体系的な改善を扱っている．これには，効果的及び効率的なマネジメントシステムの計画，実施，分析，評価及び改善が含まれる．

組織の成功に影響を及ぼす要因は，長年の間，断続的に出現，進展，増大又は消滅してきたし，こうした変化への適応が持続的成功にとって重要である．例えば，効率，品質，迅速性などこれまでに検討されていたであろうものに加えて，社会的責任，環境要因及び文化的要因が挙げられる．こうした要因は，一緒になって，組織の状況の一部となる．

持続的成功を達成する能力は，全ての階層の管理者が組織の進展する状況について学び，理解することによって強化される．改善及び革新もまた，持続的成功を支援する．

This document promotes self-assessment and provides a self-assessment tool for reviewing the extent to which the organization has adopted the concepts in this document (see **Annex A**).

A representation of the structure of this document, incorporating the elements essential to achieve sustained success of an organization as covered in this document, is presented in **Figure 1**.

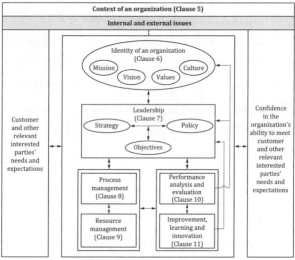

Figure 1 — **Representation of the structure of this document**

この規格では自己評価を推奨しており，組織がこの規格での概念を採用している程度をレビューするための自己評価ツールを提供している（**附属書A参照**）．

この規格が扱っている，組織が持続的成功を達成するために不可欠な要素を組み込んだ，この規格の構造図を，**図1**に示す．

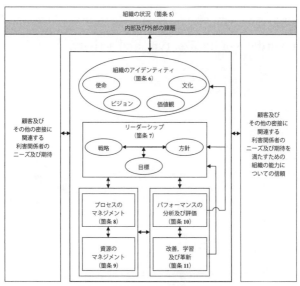

図1－この規格の構造の図示

1 Scope

This document gives guidelines for enhancing an organization's ability to achieve sustained success. This guidance is consistent with the quality management principles given in ISO 9000:2015.

This document provides a self-assessment tool to review the extent to which the organization has adopted the concepts in this document.

This document is applicable to any organization, regardless of its size, type and activity.

2 Normative references

The following documents are referred to in the text

1 適用範囲

この規格は,組織が持続的成功を達成する能力を高めるための指針を提供している.この規格は,**JIS Q 9000**:2015 に示されている品質マネジメントの原則と整合している.

この規格は,組織がこの規格の概念を採用した程度をレビューするための自己評価ツールを提供している.

この規格は,組織の規模,業種及び形態並びに活動を問わず,あらゆる組織に適用可能である.

> **注記** この規格の対応国際規格及びその対応の程度を表す記号を,次に示す.
> **ISO 9004**:2018, Quality management—Quality of an organization—Guidance to achieve sustained success (IDT)
> なお,対応の程度を表す記号"IDT"は,**ISO/IEC Guide 21-1** に基づき,"一致している"ことを示す.

2 引用規格

次に掲げる規格は,この規格に引用されることに

in such a way that some or all of their content constitutes requirements of this document. For dated references, only the edition cited applies. For undated references, the latest edition of the referenced document (including any amendments) applies.

ISO 9000:2015, *Quality management systems — Fundamentals and vocabulary*

3 Terms and definitions

For the purposes of this document, the terms and definitions given in ISO 9000:2015 apply.

ISO and IEC maintain terminological databases for use in standardization at the following addresses:

— ISO Online browsing platform: available at **https://www.iso.org/obp**

— IEC Electropedia: available at **http://www.**

よって，この規格の規定の一部を構成する．この引用規格は，記載の年の版を適用し，その後の改正版(追補を含む．)は適用しない．

JIS Q 9000:2015　品質マネジメントシステム－基本及び用語

> **注記**　対応国際規格：**ISO 9000**:2015, Quality management systems—Fundamentals and vocabulary

3 用語及び定義

この規格で用いる主な用語及び定義は，**JIS Q 9000**:2015 による．

ISO 及び **IEC** は，次の URL において，標準化に用いる用語データベースを維持する．

— **ISO** Online browsing platform：http://www.iso.org/obp
— **IEC** Electropedia：http://www.electropedia.

electropedia.org/

4 Quality of an organization and sustained success

4.1 Quality of an organization

The quality of an organization is the degree to which the inherent characteristics of the organization fulfil the needs and expectations of its customers and other interested parties, in order to achieve sustained success. It is up to the organization to determine what is relevant to achieve sustained success.

NOTE 1 The term "quality of an organization" is derived from the definition of "quality" given in ISO 9000:2015, 3.6.2 ("the degree to which a set of inherent characteristics of an object fulfils requirements"), and from the definition of "requirement" given in ISO 9000:2015, 3.6.4, ("needs or expectations that are stated, generally implied or obligatory"). It is distinct from the purpose of ISO 9001, which focuses on the quality of products and services in order to give confidence in the ability of an organization to provide conforming products

org/

4 組織の品質及び持続的成功

4.1 組織の品質

組織の品質とは,持続的成功を達成するために,組織固有の特性がその顧客及びその他の利害関係者のニーズ及び期待を満たす程度である.何が持続的成功の達成に関連しているのかの決定は,その組織に任されている.

> **注記1** "組織の品質"という用語は,**JIS Q 9000**:2015 の **3.6.2** に規定されている "品質" の定義("対象に本来備わっている特性の集まりが,要求事項を満たす程度")及び同 **3.6.4** に規定されている "要求事項" の定義("明示されている,通常暗黙のうちに了解されている又は義務として要求されている,ニーズ又は期待")から導かれている.これは,適合した製品及びサービスを提供し,顧客満足を向上すると

and services and to enhance its customers' satisfaction.

NOTE 2 All references to "needs and expectations" mean "relevant needs and expectations".

NOTE 3 All references to "interested parties" mean "relevant interested parties".

NOTE 4 All references to "interested parties" include customers.

The organization should go beyond the quality of its products and services and the needs and expectations of its customers. To achieve sustained success, the organization should focus on anticipating and meeting the needs and expectations of its interested parties, with the intent of enhancing their satisfaction and overall experience.

The organization should apply all of the quality management principles (see ISO 9000:2015) to achieve sustained success. Particular attention

いう組織の能力に信頼を与えるために，製品及びサービスの品質に重点を置く，**JIS Q 9001** の目的とは性質が異なっている．

注記 2 この規格で"ニーズ及び期待"と記載しているものは全て，"関連するニーズ及び期待"を意味する．

注記 3 この規格で"利害関係者"と記載しているものは全て，"密接に関連する利害関係者"を意味する．

注記 4 この規格で"利害関係者"と記載しているものには全て，顧客が含まれる．

組織は，その製品及びサービスの品質並びにその顧客のニーズ及び期待という範囲を超えることが望ましい．持続的成功を達成するには，組織は，利害関係者の満足度及び全体的な経験を向上させる意図をもって，それらのニーズ及び期待を予測し，満たすことに重点を置くことが望ましい．

組織は，持続的成功を達成するため，品質マネジメントの原則（**JIS Q 9000**:2015 参照）の全てを適用することが望ましい．利害関係者の様々なニー

should be given to the principles of "customer focus" and "relationship management" to meet the different needs and expectations of interested parties.

The needs and expectations of individual interested parties can be different, aligned to, or in conflict with those of other interested parties, or can change quickly. The means by which the needs and expectations of interested parties are expressed and met can take a wide variety of forms, for example co-operation, negotiation, outsourcing, or by terminating an activity; consequently, the organization should give consideration to the interrelationships of its interested parties when addressing their needs and expectations.

The composition of interested parties can differ significantly over time and between organizations, industries, cultures and nations; **Figure 2** provides examples of interested parties and their needs and expectations.

ズ及び期待を満たすため，"顧客重視"及び"関係性管理"の原則に特別な注意を払うことが望ましい．

個々の利害関係者のニーズ及び期待は異なり，その他の利害関係者のニーズ及び期待と一致する若しくは対立する，又は急速に変化する可能性がある．利害関係者のニーズ及び期待を表現し，充足する手段には，例えば，協力，交渉，外部委託又は活動の停止を含む，幅広い形がある．したがって，組織は，そのニーズ及び期待に取り組む場合には，それら利害関係者の相互関係を考慮することが望ましい．

利害関係者の構成は，時代とともに，また，組織，業種，文化及び国家によって大きく異なり得る．利害関係者並びにそのニーズ及び期待の例を，**図2**に示す．

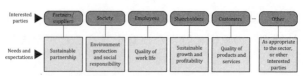

Figure 2 — **Examples of interested parties and their needs and expectations**

4.2 Managing for the sustained success of an organization

4.2.1 The quality of an organization is enhanced and sustained success can be achieved by consistently meeting the needs and expectations of its interested parties over the long term. Short- and medium-term objectives should support this long-term strategy.

As the context of an organization will be ever-changing, to achieve sustained success top management should:

a) regularly monitor, analyse, evaluate and review the organization's context in order to identify all interested parties, determine their needs and expectations and their individual potential impacts on the organization's performance;

b) determine, implement and communicate the

図2－利害関係者並びにそのニーズ及び期待の例

4.2 組織の持続的成功のためのマネジメント

4.2.1 長期にわたり，その利害関係者のニーズ及び期待を整合して満たすことによって，組織の品質が向上し，持続的成功を達成することができる．短期的目標及び中期的目標が，この長期的戦略を支援することになるのが望ましい．

組織の状況は刻々と変化するため，トップマネジメントは，持続的成功を達成するために，次の事項を行うことが望ましい．

a) 全ての利害関係者を特定し，そのニーズ及び期待並びに組織のパフォーマンスに対する個々の利害関係者の潜在的な影響を明確にするため，定期的に組織の状況を監視し，分析し，評価し，レビューする．

b) 組織の使命，ビジョン及び価値観を明確にし，

organization's mission, vision and values, and promote an aligned culture;

c) determine short- and long-term risks and opportunities;

d) determine, implement and communicate the organization's policies, strategy and objectives;

e) determine the relevant processes and manage them so that they function within a coherent system;

f) manage the organization's resources to enable its processes to achieve their intended results;

g) monitor, analyse, evaluate and review the organization's performance;

h) establish a process for improvement, learning and innovation in order to support the organization's ability to respond to changes in the context of the organization.

4.2.2 Consideration of the needs and expectations of interested parties can enable the organization:

a) to achieve objectives effectively and efficiently;

4 組織の品質及び持続的成功

実行し，伝達し，一貫性のある文化を促進する．

c) 短期的及び長期的なリスク及び機会を明確にする．

d) 組織の方針，戦略及び目標を明確にし，実施し，伝達する．

e) 論理的に首尾一貫したシステムの内部で機能するよう，関連するプロセスを明確にし，マネジメントする．

f) 組織のプロセスが意図した結果を達成することができるように，組織の資源をマネジメントする．

g) 組織のパフォーマンスを監視し，分析し，評価し，レビューする．

h) 組織の状況における変化に対応する組織の能力を支援するため，改善及び学習を行い，革新するプロセスを確立する．

4.2.2 利害関係者のニーズ及び期待を考慮することで，組織は次の事項を行えるようになる．

a) 目標を効果的及び効率的に達成する．

b) to eliminate conflicting responsibilities and relationships;
c) to harmonize and optimize practices;
d) to create consistency;
e) to improve communication;
f) to facilitate training, learning and personal development;
g) to facilitate focus on the most important characteristics of the organization;
h) to manage risks and opportunities to its brand or reputation;
i) to acquire and share knowledge.

5 Context of an organization
5.1 General

Understanding the context of an organization is a process that determines factors which influence the organization's ability to achieve sustained success. There are key factors to consider when determining the context of an organization:

a) interested parties;
b) external issues;
c) internal issues.

b) 責任及び関係の対立を排除する.

c) 実践を調和し,最適化する.
d) 整合性を生み出す.
e) コミュニケーションを改善する.
f) 訓練,学習及び個々人の能力開発を容易にする.
g) 組織の最も重要な特性への注力を容易にする.

h) ブランド又は評判に対するリスク及び機会をマネジメントする.
i) 知識を習得及び共有する.

5 組織の状況
5.1 一般

　組織の状況についての理解とは,その組織が持続的成功を達成する能力に影響を及ぼす要因を明確にするプロセスをいう.組織の状況を明確にする際に考慮すべき重要な要因には,次のものがある.

a) 利害関係者
b) 外部の課題
c) 内部の課題

5.2 Relevant interested parties

Interested parties are those that can affect, be affected by, or perceive themselves to be affected by a decision or activity of the organization. The organization should determine which interested parties are relevant. These relevant interested parties can be both external and internal, including customers, and can impact the organization's ability to achieve sustained success.

The organization should determine which interested parties:
a) are a risk to its sustained success if their relevant needs and expectations are not met;
b) can provide opportunities to enhance its sustained success.

Once the relevant interested parties are determined, the organization should:
— identify their relevant needs and expectations, determining the ones that should be addressed;
— establish the necessary processes to fulfil the needs and expectations of the interested par-

5.2 密接に関連する利害関係者

利害関係者とは，組織の意思決定若しくは活動に影響を及ぼす，又はそれらから影響を受けている可能性のある若しくは自ら影響を受けていると認識している者である．組織は，どの利害関係者が密接に関連しているのかを明確にすることが望ましい．これらの密接に関連する利害関係者は，顧客を含め，外部関係者及び内部関係者である可能性があり，組織の持続的成功を達成する能力に影響を及ぼし得る．

組織は，どの利害関係者が次の事項に該当するか，明確にすることが望ましい．
a) 関連するニーズ及び期待を満たさない場合，組織の持続的成功へのリスクとなる．
b) 組織の持続的成功を強化する機会を提供できる．

密接に関連する利害関係者を明確にしたら，組織は次の事項を行うことが望ましい．
— その関連するニーズ及び期待を特定し，取り組むことが望ましい事項を明確にする．

— 利害関係者のニーズ及び期待を満たすために必要なプロセスを確立する．

ties.

The organization should consider how to establish ongoing relationships with interested parties for benefits such as improved performance, common understanding of objectives and values, and enhanced stability.

5.3 External and internal issues

5.3.1 External issues are factors that exist outside of the organization that can affect the organization's ability to achieve sustained success, such as:

a) statutory and regulatory requirements;
b) sector-specific requirements and agreements;
c) competition;
d) globalization;
e) social, economic, political and cultural factors;
f) innovations and advances in technology;
g) natural environment.

5.3.2 Internal issues are factors that exist within the organization itself that can affect the organization's ability to achieve sustained success,

組織は，例えば，パフォーマンス改善，目標及び価値観の共通理解並びに安定性向上などの便益を得るため，利害関係者との継続的な関係を確立する方法について検討することが望ましい．

5.3 外部及び内部の課題
5.3.1 外部の課題とは，組織外部に存在し，持続的成功を達成する組織の能力に影響を及ぼし得る要因であり，次のようなものがある．

a) 法令・規制要求事項
b) 分野固有の要求事項及び合意事項
c) 競争
d) グローバル化
e) 社会的，経済的，政治的及び文化的要因
f) 技術の革新及び進歩
g) 自然環境

5.3.2 内部の課題とは，組織自体の内部に存在し，持続的成功を達成する組織の能力に影響を及ぼし得る要因であり，次のようなものがある．

such as:

a) size and complexity;
b) activities and associated processes;
c) strategy;
d) type of products and services;
e) performance;
f) resources;
g) levels of competence and organizational knowledge;
h) maturity;
i) innovation.

5.3.3 When considering external and internal issues, the organization should take into account relevant information from the past, its current situation and its strategic direction.

The organization should determine which external and internal issues could result in risks to its sustained success or opportunities to enhance its sustained success.

Based on the determination of these issues, top management should decide which of these risks

a) 規模及び複雑性
b) 活動及び関連するプロセス
c) 戦略
d) 製品及びサービスの種類
e) パフォーマンス
f) 資源
g) 力量及び組織の知識のレベル

h) 成熟度
i) 革新

5.3.3 外部及び内部の課題を検討する場合，組織は過去からの関連する情報，組織の現在の状況及びその戦略的方向性を考慮することが望ましい．

組織は，どの外部及び内部の課題が組織の持続的成功へのリスク又は持続的成功を強化する機会をもたらし得るのかを明確にすることが望ましい．

これらの課題の明確化に基づき，トップマネジメントは，これらのリスク及び機会のうちのどれに取

and opportunities should be addressed and initiate the establishment, implementation and maintenance of the necessary processes.

The organization should consider how to establish, implement and maintain a process for monitoring, reviewing and evaluating external and internal issues, with consideration of any consequences to be acted on (see **7.2**).

6 Identity of an organization
6.1 General

An organization is defined by its identity and context. The identity of an organization is determined by its characteristics, based on its mission, vision, values and culture.

Mission, vision, values and culture are interdependent and the relationship between them should be recognized as dynamic.

6.2 Mission, vision, values and culture

The identity of an organization includes its:

a) mission: the organization's purpose for exist-

り組むことが望ましいかを決断し，必要なプロセスの確立，実施及び維持を開始することが望ましい．

組織は，処置をとるべきあらゆる結果について考慮しながら，外部及び内部の課題を監視し，レビューし，評価するプロセスを確立し，実施し，維持する方法について検討することが望ましい（**7.2** 参照）．

6 組織のアイデンティティ
6.1 一般

組織は，そのアイデンティティ及び状況によって定められる．組織のアイデンティティは，その使命，ビジョン，価値観及び文化に基づいて，その特性によって決定される．

使命，ビジョン，価値観及び文化は相互に依存し合っており，その間の関係を動的なものと認識することが望ましい．

6.2 使命，ビジョン，価値観及び文化

組織のアイデンティティには，次の事項が含まれる．

a) 使命：組織が存在する目的

ing;

b) vision: aspiration of what an organization would like to become;

c) values: principles and/or thinking patterns intended to play a role in shaping the organization's culture and to determine what is important to the organization, in support of the mission and vision;

d) culture: beliefs, history, ethics, observed behaviour and attitudes that are interrelated with the identity of the organization.

It is essential that the organization's culture aligns with its mission, vision, and values. Top management should ensure that the context of the organization is considered when determining its mission, vision and values. This should include an understanding of its existing culture and assessing the need to change the culture. The strategic direction of the organization and its policy should be aligned with these identity elements.

Top management should review the mission, vision, values and culture at planned intervals and

b) ビジョン:組織がどのようになりたいのかについての願望

c) 価値観:組織の文化の形成に役割を果たし,使命及びビジョンを支持しながら何が組織にとって重要なのかを明確にすることを意図する原則及び/又は思考パターン

d) 文化:組織のアイデンティティと相互に関連する,信念,歴史,倫理,観察される行動及び態度

組織の文化が,その使命,ビジョン及び価値観と一貫していることが不可欠である.トップマネジメントは,その使命,ビジョン及び価値観を明確にする際に,組織の状況が考慮されていることを確実にすることが望ましい.これには,その既存の文化の理解,及び文化を変化させる必要性についての評価が含まれることが望ましい.組織の戦略的方向性及びその方針は,こうしたアイデンティティの要素と一貫していることが望ましい.

トップマネジメントは,計画された間隔で,また,組織の状況が変化した場合には常に,使命,ビ

whenever the context of the organization changes. This review should consider external and internal issues that can have an effect on the organization's ability to achieve sustained success. When changes are made to any of the identity elements, they should be communicated within the organization and to interested parties, as appropriate.

7 Leadership
7.1 General

7.1.1 Top management, through its leadership, should:

a) promote the adoption of the mission, vision, values and culture in a way that is concise and easy to understand, to achieve unity of purpose;

b) create an internal environment in which people are engaged and committed to the achievement of the organization's objectives;

c) encourage and support managers at appropriate levels to promote and maintain the unity of purpose and direction as established by the top management.

7.1.2 To achieve sustained success, top manage-

ジョン,価値観及び文化をレビューすることが望ましい.このレビューでは,持続的成功を達成する組織の能力に影響を及ぼす可能性がある外部及び内部の課題を考慮することが望ましい.アイデンティティの要素のいずれかに対して変更があった場合は,必要に応じて,組織内で,また,利害関係者にその変更を伝達することが望ましい.

7 リーダーシップ

7.1 一般

7.1.1 トップマネジメントは,そのリーダーシップを通して,次の事項を行うことが望ましい.

a) 簡潔かつ容易な方法で,使命,ビジョン,価値観及び文化の採用を促進し,目的の統一を図る.

b) 人々が組織の目標の達成に積極的に参画し,コミットメントする内部環境を生み出す.

c) トップマネジメントが確立したとおりに,目的及び方向性の統一を促進し,維持するよう,適切な階層の管理者を励まし,支援する.

7.1.2 持続的成功を達成するために,トップマネジ

ment should demonstrate leadership and commitment within the organization, by:

a) establishing the organization's identity (see **Clause 6**);

b) promoting a culture of trust and integrity;

c) establishing and maintaining teamwork;

d) providing people with the necessary resources, training and authority to act with accountability;

e) promoting shared values, fairness and ethical behaviour so that these are sustained at all levels of the organization;

f) establishing and maintaining an organizational structure to enhance competitiveness, where applicable;

g) individually and collectively reinforcing the organization's values;

h) communicating achieved successes externally and internally, as appropriate;

i) establishing a basis for effective communication with people in the organization, discussing issues that have general impact, including financial impact, where applicable;

j) supporting leadership development at all lev-

7 リーダーシップ

メントは次の事項によって組織内部でのリーダーシップ及びコミットメントを実証することが望ましい.

a) 組織のアイデンティティの確立（箇条 6 参照）

b) 信頼及び誠実の文化の促進
c) チームワークの確立及び維持
d) 説明責任を果たしながら行動するために必要な資源，訓練及び権限の，人々への提供

e) 共有された価値観，公平性及び倫理的行動を促進し，これらが組織の全ての階層において持続するようにすること
f) 該当する場合には，必ず，競争力を向上させる組織構造を確立し，維持すること

g) 組織の価値観の個人的及び集団的な補強

h) 必要に応じて，外部及び内部で達成された成功の伝達
i) 該当する場合には，必ず，財務影響を含む，全般的な影響を及ぼす課題を議論し，組織内の人々との効果的なコミュニケーションのための基礎を確立すること
j) 組織の全ての階層でのリーダーシップ育成支援

els of the organization.

7.2 Policy and strategy

Top management should set out the organization's intentions and direction in the form of the organization's policy, to address aspects such as compliance, quality, environment, energy, employment, occupational health and safety, quality of work life, innovation, security, privacy, data protection and customer experience. Policy statements should include commitments to satisfy the needs and expectations of interested parties and to promote improvement.

When establishing the strategy, top management should either apply a recognized and appropriate model available in the market, or design or implement an organization-specific customized model. Once chosen, it is crucial to maintain the stability of the model as the solid foundation and reference for managing the organization.

Strategy should reflect the identity of the organization, its context and long-term perspective.

7.2 方針及び戦略

　トップマネジメントは，例えば，コンプライアンス，品質，環境，エネルギー，雇用，労働安全衛生，ワークライフの質，革新，セキュリティ，プライバシー，データ保護，顧客経験などの側面に取り組むため，組織の方針という形で組織の意図及び方向性を提示することが望ましい．方針書には，利害関係者のニーズと期待を満たし，改善を促すというコミットメントを含めることが望ましい．

　戦略を定める場合には，トップマネジメントは，一般に利用可能な，認知されている適切なモデルを適用するか，又は組織固有のカスタマイズされたモデルを設計する若しくは実行することが望ましい．一旦選択したら，組織をマネジメントするための強固な基盤及び参照として，モデルの安定性を維持することが極めて重要である．

　戦略は，組織のアイデンティティ，組織の状況及び長期的な展望を反映することが望ましい．それに

All short- and medium-term objectives should be aligned accordingly (see **7.3**).

Top management should make strategic decisions regarding competitive factors (see **Table 1**).

These policy and strategy decisions should be reviewed for continued suitability. Any changes to the external and internal issues, as well as any new risks and opportunities, should be addressed.

The organization's policies and strategy constitute the basis to establish process management (see **Clause 8**).

Table 1 — Examples of actions to consider when addressing competitive factors

Competitive factors	Actions to consider
A **Products and services**	— focusing on current and potential customers and potential markets for products and services — offering standard products and services or designs specific to customer requirements — realizing the advantages of being first to market or being a follower — scaling production from one-off to mass production, as appropriate — dealing with short innovation cycles or a stable long-term customer demand — managing quality requirements

7 リーダーシップ

従って,全ての短期的及び中期的目標を一貫性のあるものにすることが望ましい(**7.3** 参照).

トップマネジメントは,競争的要因に関して戦略的な決定を行うことが望ましい(**表1**参照).

これらの方針及び戦略に関わる決定を,継続的な適切性のためにレビューすることが望ましい.外部及び内部の課題についてのあらゆる変化並びにあらゆる新しいリスク及び機会に取り組むことが望ましい.

組織の方針及び戦略は,プロセスのマネジメントを確立するための基礎となる(箇条8参照).

表1−競争的要因に取り組む場合に考慮すべき処置の例

競争要因	考慮すべき処置
A 製品及びサービス	— 現在及び潜在的な顧客,並びに製品及びサービスの潜在的な市場に焦点を当てる. — 標準的な製品及びサービス,又は顧客要求事項に対する固有の設計を提供する. — 市場の一番乗りとなる利点又はフォロワーとなる利点を実現する. — 必要に応じて,個別生産から大量生産まで生産規模を拡大・縮小する. — 短い革新サイクル又は安定した長期の顧客需要へ対処する. — 品質要求事項をマネジメントする.

Table 1 *(continued)*

Competitive factors	Actions to consider
B People	— recognizing demographic development and changing values — considering diversity — cultivating an image as an attractive employer — determining the desired competence and experience of people to hire — considering appropriate approaches to recruitment, competence development, retention, and leave management — addressing capacity flexibility by considering permanent versus fixed-term contracts — considering full time versus part time or temporary employment, as well as the balance between them
C Organizational knowledge and technology	— applying currently available knowledge and technology to new opportunities — identifying the need for new knowledge and technology — determining when this knowledge and technology needs to be available and how to apply it within the organization — determining if this should be developed internally or acquired externally
D Partners	— determining potential partners — driving joint technology development with external providers and competitors — developing customized products and services in joint undertakings with customers — co-operating with the local community, academia, public authorities and associations

7 リーダーシップ

表1（続き）

競争要因	考慮すべき処置
B 人々	— 人口増加及び価値観の変化を認識する. — 多様性を考慮する. — 魅力的な雇用者としてのイメージを養成する. — 雇用する人々に望まれる力量及び経験を明確にする. — 採用, 能力開発, 定着及び退職のマネジメントに対する適切なアプローチを考慮する. — 無期契約にするか有期契約にするかを考慮することによって, 量的能力の柔軟性へ取り組む. — フルタイムにするか, パートタイム又は臨時雇用にするかを考えるとともに, それらのバランスを考慮する.
C 組織の知識及び技術	— 新しい機会への現在利用できる知識及び技術を適用する. — 新しい知識及び技術へのニーズを特定する. — こうした知識及び技術を組織内でいつ利用可能にする必要があるのか, 並びにどのようにそれを適用するのかを決定する. — これを内部で開発するのか又は外部から獲得することが望ましいのかを決定する.
D パートナ	— 潜在的なパートナを明確にする. — 外部提供者及び競合他社との共同技術開発を推進する. — 顧客との共同事業での, カスタマイズされた製品及びサービスを開発する. — 地域社会, 学会, 公共機関及び協会と協力する.

Table 1 *(continued)*

Competitive factors	Actions to consider
E Processes	— deciding whether process management will be centralized, decentralized, integrated or non-integrated, or a hybrid approach regarding determination, establishment, maintenance, control and improvement of processes, including the assignment of roles and responsibilities — determining necessary information and communications technology (ICT) infrastructure (e.g. proprietary, customized or standard solutions)
F Place	— considering local, regional and global presence — considering virtual presence and use of social media — considering the use of virtual decentralized project teams
G Pricing	— establishing price position (e.g. low versus premium pricing strategy) — determining prices by using an auction/bidding approach

7.3 Objectives

Top management should demonstrate leadership in the organization by defining and maintaining the organization's objectives based on its policies and strategy, as well as by deploying the objectives at relevant functions, levels and processes.

Objectives should be defined for the short and long term and should be clearly understandable. Objectives should be quantified where possible. When de-

表1（続き）

競争要因	考慮すべき処置
E　プロセス	— 役割及び責任の付与を含む，プロセスの決定，確立，維持，管理及び改善に関して，プロセスのマネジメントを，集中とするのか分散とするのか，統合とするのか非統合とするのか，又はハイブリッドアプローチとするのかについて意思決定する． — 必要な情報通信技術（ICT）インフラストラクチャを決定する（例えば，専有，カスタマイズ又は標準ソリューション）．
F　場所	— 地方，地域及び世界でのプレゼンスを考慮する． — バーチャルプレゼンス及びソーシャルメディアの利用を考慮する． — 仮想分散プロジェクトチームの活用を考慮する．
G　価格設定	— 価格位置を確立する（例えば，高価格戦略か，低価格戦略か）． — 競売・入札手法の活用によって価格を決定する．

7.3　目標

トップマネジメントは，組織の方針及び戦略に基づいて組織の目標を定めて維持し，更にその目標を関連する部門，階層及びプロセスに展開することによって，組織でのリーダーシップを発揮することが望ましい．

目標は，短期的及び長期的に定め，明確に理解できるものとすることが望ましい．目標は，可能な場合，定量化することが望ましい．目標を定める場

termining the objectives, top management should consider:

a) to what extent the organization is aiming to be recognized by interested parties as:
   1) a leader with respect to competitive factors (see **7.2**) emphasizing the organization's capability;
   2) having a positive impact on economic, environmental and social conditions around it;
b) the degree of the organization's and its people's engagement in society beyond immediate business-related topics (e.g. in national and international organizations, such as public administration, associations and standardization bodies).

When deploying the objectives, top management should encourage discussions for alignment between different functions and levels of the organization.

7.4 Communication

The effective communication of policies and strategy, with relevant objectives, is essential to support

合，トップマネジメントは，次の事項を考慮することが望ましい．

a) 組織が，次のような存在として利害関係者から認識されるよう目指している程度

 1) 組織の実現能力を重視する，競争的要因（**7.2** 参照）に関するリーダー

 2) 組織を取り巻く経済的，環境的及び社会的な条件に対して，良い影響をもたらしている者

b) 直近の事業に関連するテーマを超えた，組織及びその人々の社会への積極的参加の程度（例えば，行政機関，協会，標準化団体のような，国内組織，国際組織など）

　目標を展開する場合，トップマネジメントは，組織の様々な部門と階層との間でのすり合わせのための議論を奨励することが望ましい．

7.4　コミュニケーション

　関連する目標とともに，戦略及び方針に関する効果的なコミュニケーションは，組織の持続的成功を

the sustained success of the organization.

Such communication should be meaningful, timely and continual. Communication should include a feedback mechanism and incorporate provisions to proactively address changes in the organization's context.

The organization's communication process should operate both vertically and horizontally and should be tailored to the differing needs of its recipients. For example, the same information can be conveyed in one way to people within the organization and in a different way to interested parties.

8 Process management
8.1 General
Organizations deliver value through activities con-

支援する上で不可欠である．

　このようなコミュニケーションは，有意義で，時宜を得て，継続的に行うことが望ましい．コミュニケーションには，フィードバックの仕組みを含めることが望ましく，組織の状況の変化に積極的に取り組むための備えを取り入れることが望ましい．

　組織のコミュニケーションプロセスは，垂直と水平との両方で機能し，その受け取り側の異なるニーズに合わせることが望ましい．例えば，同じ情報を，組織内の人々に対して一つの方法で，利害関係者に対して異なる方法で伝えることが可能である．

> **注記**　方針及び戦略を決定し，目標を定めて展開するためのより詳細な指針を定めた規格として，**JIS Q 9023** がある．**JIS Q 9023** は，**8.4.3** 及び **10.3～10.4** に関する，より詳細な指針も含んでいる．

8　プロセスのマネジメント
8.1　一般
　組織は，プロセスのネットワークの中で相互につ

nected within a network of processes. Processes often cross boundaries of functions within the organization. Consistent and predictable results are achieved more effectively and efficiently when the network of processes functions as a coherent system.

Processes are specific to an organization and vary depending on the type, size and level of maturity of the organization. The activities within each process should be determined and adapted to the size and distinctive features of the organization.

In order to achieve its objectives, the organization should ensure that all its processes are managed proactively, including externally provided processes, to ensure that they are effective and efficient. It is important to optimize the balance between the different purposes and specific objectives of the processes, in alignment with the organization's objectives.

This can be facilitated by adopting a "process approach" that includes establishing processes, in-

8 プロセスのマネジメント

ながっている活動を通じて価値を提供する．プロセスは，多くの場合，組織内の部門の境界をまたいでいる．プロセスのネットワークが論理的に首尾一貫したシステムとして機能している場合には，整合性があり，予測可能な結果が，より効果的及び効率的に達成されている．

プロセスはそれぞれの組織に固有のものであり，組織の業種及び形態，規模並びに成熟度によって異なる．各プロセス内の活動は，組織の規模及び顕著な特徴に応じて決定し，適応させることが望ましい．

組織は，その目標を達成するために，外部から提供されるプロセスを含む，全てのプロセスが効果的かつ効率的であることを確実にするよう，それらのプロセスを積極的にマネジメントすることが望ましい．組織の目標との一貫性を保ちながら，プロセスの様々な目的と特定の目標との間のバランスを最適化することが重要である．

これは，プロセス，相互依存性，制約条件及び資源配分を確立することを含む"プロセスアプロー

terdependencies, constraints and shared resources.

NOTE For further information on the "process approach", see the related quality management principles in ISO 9000:2015, and the "ISO 9001:2008 Introduction and Support Package" document *Guidance on the concept, content and use of the process approach for management systems,* available from: **https://committee.iso.org/tc176sc2**.

8.2 Determination of processes

8.2.1 The organization should determine the processes and their interactions necessary for providing outputs that meet the needs and expectations of interested parties, on an ongoing basis. **Figure 3** gives a schematic representation of a process.

チ"を採用することによって,容易に行うことができる.

> **注記** "プロセスアプローチ"については,**JIS Q 9000**:2015 の関連する品質マネジメントの原則,及び"**ISO 9001**:2008 導入・支援パッケージ"文書－ Guidance on the Concept and Use of the Process Approach for management systems を参照."**ISO 9000** 導入・支援"文書は,次の URL で提供されている.https://committee.iso.org/tc176sc2

8.2 プロセスの決定

8.2.1 組織は,利害関係者のニーズ及び期待を満たすアウトプットを継続的に提供するために必要となる,プロセス及びその相互作用を決定することが望ましい.単一プロセスの略図を,**図 3** に示す.

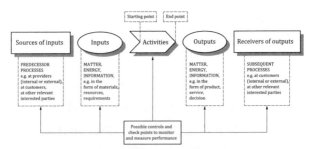

Figure 3 — Schematic representation of the elements of a single process

Processes and their interactions should be determined in accordance with the organization's policies, strategy and objectives, and should address areas such as:

a) operations related to products and services;

b) meeting the needs and expectations of interested parties;

c) the provision of resources;

d) managerial activities, including monitoring, measuring, analysis, review, improvement, learning and innovation.

8.2.2 In determining its processes and their interactions, the organization should give consideration, as appropriate, to:

8 プロセスのマネジメント

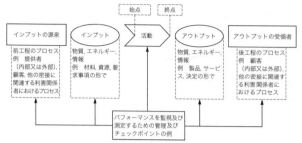

図3－単一プロセスの要素の略図

プロセス及びその相互作用は，組織の方針，戦略及び目標に従って決定し，次のような領域を扱うものにすることが望ましい．

a) 製品及びサービスに関連する運用
b) 利害関係者のニーズ及び期待の充足

c) 資源の提供
d) 監視，測定，分析，レビュー，改善，学習及び革新を含むマネジメント活動

8.2.2 プロセス及びその相互作用の決定においては，必要に応じて，組織は，次の事項を考慮することが望ましい．

a) the purpose of the process;

b) the objectives to be achieved and related performance indicators;

c) the outputs to be provided;

d) the needs and expectations of interested parties, and their changes;

e) changes in operations, markets and technologies;

f) the impacts of the processes;

g) the inputs, resources and information needed, and their availability;

h) the activities that need to be implemented and methods that can be used;

i) constraints for the process;

j) risks and opportunities.

8.3 Responsibility and authority for processes

For each process, the organization should appoint a person or a team (often referred to as the "process owner"), depending on the nature of the process and the organization's culture, with defined responsibilities and authorities to determine, maintain, control and improve the process and its interaction with other processes it impacts and those

8 プロセスのマネジメント

a) プロセスの目的
b) 達成すべき目標及び関連するパフォーマンス指標
c) 提供すべきアウトプット
d) 利害関係者のニーズ及び期待並びにその変化
e) 運用,市場及び技術の変化
f) プロセスの影響
g) 必要となるインプット,資源及び情報並びにその利用可能性
h) 実施する必要のある活動及び使用できる方法
i) プロセスにおける制約条件
j) リスク及び機会

8.3 プロセスの責任及び権限

　組織は,各プロセスに対して,プロセスの性質及び組織の文化に応じて,個人又はチーム(しばしば"プロセスオーナ"と呼ばれる.)を任命し,プロセス及びそのプロセスが影響を及ぼし,かつ,そのプロセスに影響を与える他のプロセスとの相互作用を決定し,維持し,管理し,改善するための,責任及び権限を定めることが望ましい.組織は,プロセス

that have impact on it. The organization should ensure that the responsibilities, authorities and roles of process owners are recognized throughout the organization and that the people associated with the individual processes have the competences needed for the tasks and activities involved.

8.4 Managing processes

8.4.1 To manage its processes effectively and efficiently, the organization should:

a) manage the processes and their interactions, including externally provided processes, as a system to enhance alignment/linkage between the processes;

b) visualize the network of processes, their sequence and interactions in a graphic (e.g. process map, diagrams) in order to understand the roles of each process in the system and its effects on the performance of the system;

c) determine criteria for the outputs of processes, evaluate the capability and performance of processes by comparing the outputs with the criteria, and plan actions to improve the processes when they are not effectively achiev-

オーナの責任，権限及び役割が組織全体を通して認識されること，並びに個々のプロセスに関連する人々が関与する業務及び活動のために必要な力量をもっていることを確実にすることが望ましい．

8.4 プロセスのマネジメント

8.4.1 組織は，効果的及び効率的にプロセスをマネジメントするため，次の事項を行うのが望ましい．

a) プロセス間のすり合わせ／連携を強化するために，外部から提供されたプロセスを含め，プロセス及びその相互作用を一つのシステムとしてマネジメントする．

b) システム内における各プロセスの役割及びそのシステムのパフォーマンスへの影響を理解するため，プロセスのネットワーク，その順序及び相互作用を図によって視覚化する（例えば，プロセスマップ，ダイアグラムなど）．

c) プロセスのアウトプットに対する基準を明確にし，アウトプットと基準とを比較することによって，プロセスの実現能力及びパフォーマンスを評価し，プロセスがシステムによって想定されているパフォーマンスを効果的に達成することができな

ing the performance expected by the system;

d) assess the risks and opportunities associated with the processes and implement any actions that are necessary in order to prevent, detect and mitigate undesired events, including risks such as:

   1) human factors (e.g. shortage of knowledge and skills, rule violations, human errors);
   2) inadequate capability, deteriorations and breakdowns of equipment;
   3) design and development failure;
   4) unplanned changes in incoming materials and services;
   5) uncontrolled variation in the environment for the operation of processes;
   6) unexpected changes in the needs and expectations of interested parties, including market demand;

e) review the processes and their interrelationships on a regular basis and take suitable actions for control and improvement, to ensure they continue to be effective and support the sustained success of the organization.

い場合には,プロセスを改善する処置を計画する.
- **d)** プロセスに関連するリスク及び機会を評価し,次のようなリスクを含む,望ましくない事象を防止し,検出し,軽減するために必要な処置を実施する.

  1) 人的要因(例えば,知識及び技能の不足,規則違反,人的ミス)
  2) 設備の不十分な実現能力,劣化及び破損
  3) 設計・開発の失敗
  4) 受け入れる材料及びサービスにおける計画外の変更
  5) プロセスを運用するための環境における管理されていない変動
  6) 市場需要を含む,利害関係者のニーズ及び期待における予想外の変化

- **e)** プロセス及びその相互関係が効果的であり続け,組織の持続的成功を支援することを確実にするために,それらを定期的にレビューし,管理及び改善のための適切な処置を実施する.

8.4.2 Processes should operate together within a coherent management system. Some processes will relate to the overall management system and some will additionally relate to a specific managerial aspect, such as:

a) the quality of products and services, including cost, quantity and delivery (e.g. ISO 9001);

b) health, safety, security (e.g. ISO 45001, ISO/IEC 27001);

c) environment, energy (e.g. ISO 14001, ISO 50001);

d) social responsibility, anti-bribery, compliance (e.g. ISO 26000, ISO 37001, ISO 19600);

e) business continuity, resilience (e.g. ISO 22301, ISO 22316).

8.4.3 To attain a higher level of performance, the processes and their interactions should be continually improved according to the organization's policies, strategy and objectives (see **7.2** and **7.3**), including consideration of the need to develop or acquire new technologies, or to develop new products and services or their features, for added value.

8 プロセスのマネジメント

8.4.2 各プロセスは，論理的に首尾一貫したマネジメントシステムの内部で，一体のものとして運用することが望ましい．マネジメントシステム全体に関係するプロセスもあれば，次のような特定のマネジメント側面に関係するプロセスもある．

a) コスト，数量及び納期を含む製品及びサービスの品質（**例　JIS Q 9001**）

b) 労働安全衛生，情報セキュリティ（**例　JIS Q 45001，JIS Q 27001**）

c) 環境，エネルギー（**例　JIS Q 14001，JIS Q 50001**）

d) 社会的責任，反贈賄，コンプライアンス（**例　JIS Z 26000，ISO 37001，ISO 19600**）

e) 事業継続，レジリエンス（**例　JIS Q 22301，ISO 22316**）

8.4.3 より高いパフォーマンスを達成するために，組織の方針，戦略及び目標（**7.2** 及び **7.3** 参照）に従って，新しい技術を開発若しくは獲得する必要性，又は付加価値を付ける新しい製品及びサービス若しくはその特徴を開発する必要性を考慮することを含め，プロセス及びその相互作用を継続的に改善することが望ましい．

The organization should motivate people to engage in improvement activities and propose opportunities for improvement in the processes in which they are involved.

The organization should regularly review the achievement of objectives for the improvement of processes and their interactions, the progress of action plans, and the effects on the organization's policies, objectives and strategies. It should take any necessary corrective actions, or other appropriate actions, when gaps are identified between the planned and actual activities.

8.4.4 To maintain the level of performance attained, processes should be operated under controlled conditions, regardless of any planned and unplanned changes. The organization should determine what procedures (if any) are needed to manage a process, including the criteria for process outputs and operational conditions, to ensure conformity with the criteria.

When procedures are applied, in order to ensure

組織は，人々が改善活動に積極的に参加し，自分が関わっているプロセスにおいて改善の機会を提案するよう動機付けることが望ましい．

組織は，プロセス及びその相互作用に関する改善目標の達成，実施計画の進捗状況並びに組織の方針，目標及び戦略への影響について定期的にレビューすることが望ましい．計画した活動と実際の活動との間にギャップを特定した場合，組織は，必要な是正処置又はその他の適切な処置をとることが望ましい．

8.4.4 達成されたパフォーマンスのレベルを維持するため，計画された及び計画外の変更に関係なく，管理された条件の下でプロセスを運営することが望ましい．組織は，基準との適合を確実にするため，プロセスのアウトプット及び運用条件に対する基準を含め，（もしあるとすれば）どんな手順がプロセスをマネジメントするために必要となるのかを明確にすることが望ましい．

組織は，手順を適用する場合，プロセスの運用に

that they are followed by people involved in the operation of the process, the organization should ensure that:

a) a system is established to define the knowledge and skills needed for processes and evaluate the knowledge and skills of the process operators;

b) risks in the procedures are identified, assessed and reduced by improving the procedures (e.g. making it difficult to make errors or not allowing progression to the next process if an error occurs);

c) resources necessary for people to follow the procedures are made available;

d) people have the knowledge and skills needed for following the procedures;

e) people understand the impacts of not following the procedures (e.g. by using examples of experienced undesired events) and managers at appropriate levels take the actions that are necessary whenever a procedure is not followed;

f) consideration is given to learning, training, motivation and prevention of human error.

関わる人々がそれに従っていることを確実にするため，次の事項を確実にすることが望ましい．

a) プロセスに必要な知識及び技能を定め，プロセスを運用する人の知識及び技能を評価するためのシステムを確立している．

b) 手順におけるリスクを特定し，評価し，その手順を改善することによって低減される（例えば，誤りを犯しにくくなる，又は誤りが発生したら次のプロセスに進行できなくなる．）．

c) 人々が手順に従うために必要な資源を提供している．

d) 人々が手順に従うために必要な知識及び技能を備えている．

e) 人々は，（例えば，経験したことのある望ましくない事象の例を用いることによって）手順に従わないことによる影響を理解しており，手順に従っていない場合には，常に，適切な階層の管理者が必要な処置をとっている．

f) 教育，訓練，動機付け及び人的ミスの防止への配慮が行われている．

8.4.5 The organization should monitor its processes on a regular basis to detect deviations, and should identify and take appropriate actions when necessary without delay. Deviations are mainly caused by changes in equipment, method, material, measurement, environment and people for the operation of processes. The organization should determine check points and related performance indicators that will be effective and efficient in detecting deviations.

9 Resource management
9.1 General
Resources support the operation of all processes in

8.4.5 組織は，定期的にそのプロセスを監視して，必要な場合には遅滞なく逸脱を検出し，適切な処置を特定し，実施することが望ましい．逸脱は，主にプロセス運用のための設備，方法，材料，測定，環境及び人々における変化によって引き起こされる．組織は，逸脱を検出するために効果的かつ効率的であるチェックポイント及び関係するパフォーマンス指標を明確にすることが望ましい．

> 注記1　プロセスを効果的かつ効率的にマネジメントし，首尾一貫したマネジメントシステムの内部で，一体のものとして運用するためのより詳細な指針を定めた規格として，**JIS Q 9027** がある．
>
> 注記2　達成したプロセスのパフォーマンスのレベルを維持するためのより詳細な指針を定めた規格として，**JIS Q 9026** がある．

9 資源のマネジメント

9.1 一般

資源は，組織における全てのプロセスの運用を支

an organization and are critical for ensuring effective and efficient performance and its sustained success.

The organization should determine and manage the resources needed for the achievement of its objectives, taking into account the associated risks and opportunities and their potential effects.

Examples of key resources include:

a) financial resources;
b) people;
c) organizational knowledge;
d) technology;
e) infrastructure, such as equipment, facilities, energy and utilities;
f) the environment for the organization's processes;
g) the materials needed for the provision of products and services;
h) information;
i) resources provided externally, including subsidiaries, partnerships and alliances;

援し,効果的及び効率的なパフォーマンス並びにその持続的成功を確実にするために必要不可欠である.

組織は,関連するリスク及び機会並びにそれらの潜在的な影響を考慮しながら,組織の目標の達成に必要な資源を明確にし,マネジメントすることが望ましい.

重要な資源の例として,次のような事項が挙げられる.
a) 財務資源
b) 人々
c) 組織の知識
d) 技術
e) 設備,施設,エネルギー,ユーティリティなどのインフラストラクチャ
f) 組織のプロセスのための環境

g) 製品及びサービスの提供に必要な材料

h) 情報
i) 子会社,パートナ及び同盟関係組織を含む,外部から提供された資源

j)   natural resources.

The organization should implement sufficient control over its processes to achieve efficient and effective use of its resources. Depending on the nature and complexity of the organization, some of the resources will have different impacts on the sustained success of the organization.

When considering future activities, the organization should take into account the accessibility and suitability of resources, including externally provided resources. The organization should frequently evaluate its existing use of resources to determine opportunities for improving their use, optimizing processes, and implementing new technologies to reduce risks.

9.2   People
9.2.1   General
Competent, engaged, empowered and motivated people are a key resource. The organization should develop and implement processes to attract and retain people who have the current or potential com-

j) 天然資源

組織は，その資源の効率的及び効果的な利用を達成するために，そのプロセスに対する十分な管理を実施することが望ましい．組織の性質及び複雑性によって，幾つかの資源は組織の持続的成功に対して様々な影響を及ぼす．

組織は，将来の活動を考える場合，外部からの提供を含む資源について，入手の可能性及び適切性を考慮することが望ましい．組織は，その利用を改善し，プロセスを最適化し，リスクを低減させる新しい技術を取り入れる機会を明確にするために，既存の資源利用を頻繁に評価することが望ましい．

9.2 人々
9.2.1 一般

力量があり，積極的に参加し，権限委譲され，動機付けられた人々は，重要な資源である．組織は，組織への十分な貢献を行うための，現在ある又は潜在的な力量及び対応力をもつ人々を引き付け，保持

petences and availability to contribute fully to the organization. The managing of people should be performed through a planned, transparent, ethical and socially responsible approach at all levels throughout the organization.

9.2.2 Engagement of people

Engagement of people enhances the organization's ability to create and deliver value for interested parties. The organization should establish and maintain processes for engagement of its people. Managers at all levels should encourage people to be involved in improving performance and meeting the organization's objectives.

To enhance the engagement of its people, the organization should consider activities such as:

a) developing a process to share knowledge;
b) making use of its people's competence;
c) establishing a skills qualification system and career planning to promote personal development;
d) continually reviewing their level of satisfaction, relevant needs and expectations;

するプロセスを開発し，実施することが望ましい．人々のマネジメントは，組織全体の全ての階層において計画的で，透明で，倫理的で，社会的責任を果たすアプローチで実施することが望ましい．

9.2.2 人々の積極的参加

人々の積極的参加は，利害関係者への価値を創造し，提供する組織の能力を強化する．組織は，組織の人々の積極的参加のためのプロセスを確立し，維持することが望ましい．全ての階層の管理者は，人々がパフォーマンスの改善及び組織の目標の充足に参画するよう奨励することが望ましい．

組織は，組織の人々の積極的参加を向上させるため，次のような活動を検討することが望ましい．
a) 知識を共有するプロセスの開発
b) 人々の力量の活用
c) 個人の能力開発を促すための技能認定制度及びキャリアプランの確立

d) 人々の満足の度合い，並びに関連するニーズ及び期待についての継続的なレビューの実施

e) providing opportunities for mentoring and coaching;

f) promoting team improvement activities.

9.2.3 Empowerment and motivation of people

Empowered and motivated people at all levels throughout the organization are essential to enhance the organization's ability to create and deliver value. Empowerment enhances the motivation of people to take responsibility for their work and its results. This can be achieved by providing people with the necessary information, authority and freedom to make decisions related to their own work. Managers at all levels should motivate people to understand the significance and importance of their responsibilities and activities in relation to creating value for interested parties. To enhance the empowerment and motivation of people, managers at all levels should:

a) define clear objectives (aligned with the organization's objectives), delegate authority and responsibility, and create a work environment in which people control their own work and decision making;

e) 指導（mentoring）及びコーチングのための機会の提供

f) チームによる改善活動の促進

9.2.3 人々への権限委譲及び動機付け

　組織の全ての階層において，権限を委譲され，動機付けされている人々は，価値を創造し，提供する組織の能力を強化するために不可欠である．権限の委譲によって，人々がその作業及び結果に対して責任をもとうとする動機が強化される．これは，人々に自らの作業に関連した決定を行うために必要な情報，権限及び自由度を付与することで達成できる．全ての階層の管理者は，人々が利害関係者に対する価値の創造及び提供に関連する，その責任及び活動の意義並びに重要性を理解するよう動機付けることが望ましい．全ての階層の管理者は，人々の権限委譲及び動機付けを強化するため，次の事項を行うことが望ましい．

a) （組織の目標と一貫性のある）明確な目標を定め，権限及び責任を委任し，人々が自らの作業及び意思決定を管理する作業環境を生み出す．

b) introduce an appropriate recognition system, based on the evaluation of people's accomplishments (both individually and in teams);

c) provide incentives for people to act with initiative (both individually and in teams), as well as recognizing good performance, rewarding results and celebrating the achievement of objectives.

9.2.4 Competence of people

A process should be established and maintained to assist the organization in determining, developing, evaluating and improving the competence of people at all levels. The process should follow steps such as:

a) determining and analysing the personal competencies needed by the organization in accordance with its identity (mission, vision, values and culture), strategy, policies and objectives;

b) determining the current competence at group level and at individual level, as well as the gaps between what is available and what is currently needed, or could be needed in the future;

c) implementing actions to improve and acquire

b) 人々の業績の評価(個人及びチームとして)に基づいた,適切な表彰制度を導入する.

c) (個人及びチーム内で)人々が率先して行動するようになるためのインセンティブを提供するとともに,優れたパフォーマンスを認め,結果に対する褒賞を与え,目標達成を祝福する.

9.2.4 人々の力量

全ての階層の人々の力量を明確にし,開発し,評価し,改善するよう組織を支援するためのプロセスを確立し,維持することが望ましい.そのプロセスは,次のようなステップに従うことが望ましい.

a) 組織のアイデンティティ(使命,ビジョン,価値観及び文化),戦略,方針及び目標に従って,組織が必要とする個人の力量を明確にし,分析する.

b) 集団及び個人のレベルでの現在の力量,並びに利用できるものと現在必要とされているもの又は今後必要となり得るものとの間のギャップを明確にする.

c) 必要に応じて,力量を改善し,獲得するための

competence, as required;

d) improving and maintaining competence that has been acquired;

e) reviewing and evaluating the effectiveness of actions taken to confirm that the necessary competence has been acquired.

9.3 Organizational knowledge

9.3.1 Organizational knowledge can be based on external or internal sources. Top management should:

a) recognize knowledge as an intellectual asset and manage it as an essential element of the organization's sustained success;

b) consider the knowledge required to support the short- and long-term needs of the organization, including succession planning;

c) assess how the organization's knowledge is identified, captured, analysed, retrieved, maintained and protected.

9.3.2 When defining how to determine, maintain and protect knowledge, the organization should develop processes to address:

処置を実施する．
d) 獲得している力量を改善し，維持する．

e) 必要な力量を獲得していることを確認するためにとった処置の有効性をレビューし，評価する．

9.3 組織の知識
9.3.1 組織の知識は，外部又は内部の情報源に基づくことが可能である．トップマネジメントは，次の事項を行うことが望ましい．
a) 知識を，知的財産として認識し，それを組織の持続的成功に不可欠な要素としてマネジメントする．
b) 伝承の計画策定を含め，組織の短期的及び長期的ニーズを支えるために必要となる知識を検討する．
c) 組織の知識を特定し，取得し，分析し，検索し，維持し，保護する方法を評価する．

9.3.2 知識を明確にし，維持し，保護する方法を定める際には，組織は，次の事項に取り組むプロセスを開発することが望ましい．

a) lessons learned from failures and successful projects;

b) explicit and tacit knowledge that exists within the organization, including the knowledge, insights and experience of its people;

c) determining the need to acquire knowledge from interested parties as part of the organization's strategy (see **9.6**);

d) confirming the effective distribution and understanding of information, throughout the life cycle(s) of the organization's products and services;

e) managing documented information and its use;

f) managing intellectual property.

9.4 Technology

Top management should consider technological developments, both existing and emerging, that can have a significant impact on the organization's performance in processes related to product and service provision, marketing, competitive advantage, agility and interaction with interested parties. The organization should implement processes

a)　失敗及び成功したプロジェクトから学んだ教訓

b)　組織の人々の知識,洞察及び経験を含む,組織内部に存在する形式知及び暗黙知

c)　組織の戦略の一部として利害関係者から知識を獲得する必要性の明確化(**9.6**参照)

d)　組織の製品及びサービスのライフサイクル全体を通じた情報の,効果的な配布及び理解の確認

e)　文書化した情報のマネジメント及びその使用

f)　知的財産のマネジメント

9.4　技術

　トップマネジメントは,製品及びサービスの提供,マーケティング,競争優位,迅速性並びに利害関係者との相互作用に関連するプロセスにおける,組織のパフォーマンスに対して,重大な影響を及ぼし得る,既存及び新興の両方の技術開発を検討することが望ましい.組織は,次の事項を検討することによって,技術開発及び革新を見出すためのプロセ

for detecting technological developments and innovations by considering:

a) the current levels, and emerging trends of technology, both within and external to the organization;
b) the financial resources needed to adopt the technological changes, or to acquire another organization's technological capabilities, and the benefits of such changes;
c) the organizational knowledge and capability to adapt to the technological changes;
d) the risks and opportunities;
e) the market environment.

9.5 Infrastructure and work environment
9.5.1 General

Infrastructure and work environment are key to the effective and efficient operation of all processes in the organization. The organization should determine what is needed and coordinate how these resources will be allocated, provided, measured or monitored, optimized, maintained and protected.

The organization should periodically evaluate the

スを実施することが望ましい．

a) 組織内外における，現在のレベル及び新興の技術動向

b) 技術的変化を適応する，又は別の組織の技術的な実現能力を獲得するために必要な財務資源及びそうした変化の便益

c) 技術変化に適用する組織の知識及び実現能力

d) リスク及び機会
e) 市場環境

9.5 インフラストラクチャ及び作業環境
9.5.1 一般

インフラストラクチャ及び作業環境は，組織における全てのプロセスの効果的及び効率的な運用にとって鍵となる．組織は，何が必要かを明確にするとともに，こうした資源を配分し，提供し，測定し，又は監視し，最適化し，維持し，保護する方法を組み合わせることが望ましい．

組織は，望ましいパフォーマンス及び組織の目標

suitability of the infrastructure and work environment of all related processes to achieve the desired performance and the organization's objectives.

9.5.2 Infrastructure

In managing its infrastructure, the organization should give appropriate consideration to factors such as:

a) dependability (including consideration of availability, reliability, maintainability and maintenance support, as applicable, including safety and security);

b) infrastructure elements needed for the provision of processes, products and services;

c) the efficiency, capacity and investment required;

d) the impact of the infrastructure.

9.5.3 Work environment

In determining a suitable work environment, the organization should give appropriate consideration to factors (or a combination of factors) such as:

a) physical characteristics such as heat, humidity, light, airflow, hygiene, cleanliness and noise;

を達成するための，全ての関連するプロセスについて，それらのインフラストラクチャ及び作業環境の適切性を，定期的に評価することが望ましい．

9.5.2 インフラストラクチャ

組織は，インフラストラクチャをマネジメントする際に，次のような要因を適切に考慮することが望ましい．

a) ディペンダビリティ（安全性及びセキュリティを含む，必要に応じて，入手可能性，信頼性，保全性及び保全支援を考慮したもの）

b) プロセス，製品及びサービスの提供に必要なインフラストラクチャの要素

c) 必要とする効率，量的能力及び投資

d) インフラストラクチャの影響

9.5.3 作業環境

組織は，適切な作業環境を明確にする際に，次のような要因（又は要因の組合せ）について適切に考慮することが望ましい．

a) 熱，湿度，明度，空気の流れ，衛生，清浄，騒音などの，物理的特性

b) ergonomically designed work stations and equipment;
c) psychological aspects;
d) encouraging personal growth, learning, knowledge transfer and teamwork;
e) creative work methods and opportunities for greater involvement, to realize the potential of people in the organization;
f) health and safety rules and guidance, as well as the use of protective equipment;
g) workplace location;
h) facilities for people in the organization;
i) optimization of resources.

The organization's work environment should encourage productivity, creativity and well-being for the people working in or visiting its premises (e.g. customers, external providers, partners). In addition, depending on its nature, the organization should verify that its work environment complies with applicable requirements and addresses applicable standards (such as those for environmental and occupational health and safety management).

9 資源のマネジメント

b) 人間工学的に設計された職場及び設備

c) 心理的側面
d) 個人の成長,学習及び知識の移転並びにチームワークの奨励
e) 組織内の人々の潜在能力を引き出せるよう,参画を高める創造的な作業の方法及び機会

f) 安全衛生に関わる規則及び手引,並びに保護具の使用
g) 職場の場所
h) 組織内の人々のための施設
i) 資源の最適化

　組織の作業環境は,組織の敷地内で作業をする人々又は組織の敷地を訪問する人々(例えば,顧客,外部提供者,パートナ)の生産性,創造性及び快適性を促進することが望ましい.また,組織は,その性質に応じて,作業環境が,適用される要求事項を遵守し,適用される基準(環境マネジメントに関するもの,労働安全衛生マネジメントに関するものなど)に対応していることを検証することが望ましい.

9.6 Externally provided resources

Organizations procure externally supplied resources from a variety of providers. As these resources can impact both the organization and its interested parties, it is essential that its relationships with external providers and partners are managed effectively. The organization and its external providers or partners are interdependent. The organization should seek to establish relationships that enhance the capabilities of itself and its providers or partners to create value in a manner that is mutually beneficial to all involved.

The organization should consider partnering if external providers have knowledge that the organization does not have, or to share the risks and opportunities associated with its projects (and the resulting profits or losses). Partners can be external providers of processes, products or services, technological and financial institutions, governmental and non-governmental organizations, or other interested parties.

The managing of external providers should take

9.6 外部から提供される資源

　組織は，様々な提供者から外部供給される資源を調達する．これらの資源は，組織及びその利害関係者の双方に影響を及ぼす可能性があることから，外部提供者及びパートナとの関係を効果的にマネジメントすることは不可欠である．組織とその外部提供者若しくはパートナとは，相互に依存している．組織は，参画する全ての者にとって互いに有益となる方法で，組織自身の価値，及びその提供者又はパートナの価値を創造する実現能力を向上させる関係が確立することを目指すことが望ましい．

　組織にはない知識を外部提供者が保有している場合，又はそのプロジェクトに関連するリスク及び機会（並びに結果として得られる利益又は損失）を共有するため，組織は，提携を検討することが望ましい．パートナとなり得るのは，プロセス，製品又はサービスの外部提供者，技術機関及び金融機関，政府及び非政府組織，又はその他の利害関係者である．

　外部提供者のマネジメントに当たっては，次の事

into account the risks and opportunities associated with:

a) internal facilities or capacity;
b) the technical capability to fulfil the requirements for products or services;
c) the availability of qualified resources;
d) the type and extent of controls needed for external providers;
e) business continuity and supply chain aspects (e.g. high dependability on a single or limited number of providers);
f) environmental, sustainability and social responsibility aspects.

In order to establish mutually beneficial relationships and to enhance the abilities of external providers and partners for managing activities, processes and systems, the organization should:

— share its mission and vision (and possibly its values and culture) with them;
— provide any necessary support (in terms of resources or knowledge).

9.7 Natural resources

項に関連するリスク及び機会を考慮に入れることが望ましい.

a) 内部の施設又は量的能力
b) 製品又はサービスの要求事項を満たす技術的な実現能力
c) 適格な資源の入手可能性
d) 外部提供者に対して必要とされる管理の種類及び範囲
e) 事業継続及びサプライチェーンの側面(例えば,単一又は限られた数の提供者への高度な依存性)
f) 環境,持続性及び社会的責任の側面

 組織は,互恵関係を確立し,外部提供者及びパートナの,活動,プロセス及びシステムをマネジメントする能力を高めるため,組織は次のことを行うことが望ましい.
— その使命及びビジョン(並びに恐らくその価値観及び文化)を外部提供者及びパートナと共有する.
— (資源又は知識の点で)必要な支援を行う.

9.7 天然資源

The organization should recognize its responsibility to society and should act based on this recognition. The responsibility includes several aspects, such as natural resources and the environment.

In terms of managing resources, the natural resources consumed by the organization in the provision of products and services are a strategic issue affecting its sustained success. The organization should address how to determine, obtain, maintain, protect and use essential resources such as water, soil, energy or raw materials.

The organization should address both the current and future use of natural resources required by its processes, as well as the impact of the use of natural resources related to the life cycle of its products and services. This should also be aligned with the organization's strategy.

Good practices for managing natural resources for sustained success include:

a) treating them as a strategic business matter;
b) being aware of new trends and technologies on

9 資源のマネジメント

組織は，その社会への責任を認識し，この認識に基づいて行動することが望ましい．その責任には，天然資源及び環境のような幾つかの側面が含まれる．

資源のマネジメントという観点から見ると，製品及びサービスの提供において，組織によって消費される天然資源は，その持続的成功に影響を及ぼす戦略的な課題である．組織は，水，土壌，エネルギー，原材料などの不可欠な資源をどのように明確にし，取得し，維持し，保護し，利用するかについて取り組むことが望ましい．

組織は，そのプロセスが必要とする天然資源の現在及び今後の双方の利用，並びに製品及びサービスのライフサイクルに関連する天然資源の利用による影響に取り組むことが望ましい．また，これは，組織の戦略と一貫性があることが望ましい．

持続的成功のための天然資源のマネジメントにおける優れた実践として，次の事項が挙げられる．
a) 天然資源を戦略的事業事項として取り扱うこと
b) 天然資源の効率的な利用及び利害関係者の期待

their efficient use, and on the expectations of interested parties;

c) monitoring their availability and determining the potential risks and opportunities on their use;

d) defining future markets, products and services and the impact on their use throughout the life cycle;

e) implementing best practices in their current application and use;

f) improving the actual use and minimizing the potential undesirable impact of their use.

10 Analysis and evaluation of an organization's performance

10.1 General

The organization should establish a systematic approach to collect, analyse and review available information. Based on the results, the organization should use the information to update its understanding of its context, policies, strategy and objectives as needed, while also promoting improvement, learning and innovation activities.

に関する，新しい傾向及び技術を認識すること

c) 天然資源の入手可能性を監視し，利用に関する潜在的なリスク及び機会を明確にすること

d) 今後の市場，製品及びサービス，並びにライフサイクル全体を通じた天然資源の利用への影響を定めること

e) 天然資源の現在の適用及び利用におけるベストプラクティスを実施すること

f) 実際の利用を改善し，天然資源の利用による潜在的な望ましくない影響を最小限に抑えること

10 組織のパフォーマンスの分析及び評価

10.1 一般

組織は，利用可能な情報を収集し，分析し，レビューする体系的なアプローチを確立することが望ましい．その結果に基づき，組織は，改善，学習及び革新活動も促進しながら，必要に応じて，組織の状況，方針，戦略及び目標についての理解を更新するために，情報を利用することが望ましい．

The available information should include data on:

a) the organization's performance (see **10.2**, **10.3** and **10.4**);
b) the status of the organization's internal activities and resources, which can be understood through internal audits or self-assessment (see **10.5** and **10.6**);
c) changes in the organization's external and internal issues and the needs and expectations of the interested parties.

10.2 Performance indicators

10.2.1 The organization should assess its progress in achieving its planned results against its mission, vision, policies, strategy and objectives, at all levels and in all relevant processes and functions. A measurement and analysis process should be used to monitor this progress, to gather and provide the information necessary for performance evaluations and effective decision making.

The selection of appropriate performance indicators and monitoring methods is critical for effec-

10 組織のパフォーマンスの分析及び評価　113

利用可能な情報には，次の事項に関するデータを含めることが望ましい．

a) 組織のパフォーマンス（**10.2**，**10.3** 及び **10.4** 参照）

b) 内部監査又は自己評価を通じて理解することができる，組織の内部活動及び資源の状態（**10.5** 及び **10.6** 参照）

c) 組織の外部及び内部の課題，並びに利害関係者のニーズ及び期待における変化

10.2　パフォーマンス指標

10.2.1　組織は，全ての階層並びに全ての関係するプロセス及び部門において，組織の使命，ビジョン，方針，戦略及び目標に照らし，計画した結果が達成できているかどうか進捗状況を評価することが望ましい．この進捗状況を監視し，パフォーマンス評価及び効果的な意思決定のために必要な情報を収集し，提供するために，測定プロセス及び分析プロセスを使用することが望ましい．

適切なパフォーマンス指標及び監視方法の選定は，組織の効果的な測定及び分析にとって必要不可

tive measurement and analysis of an organization. **Figure 4** shows steps for using performance indicators.

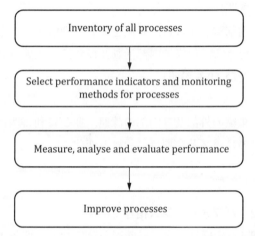

Figure 4 — Steps for using performance indicators

10.2.2 The methods used for collecting information regarding performance indicators should be practicable and appropriate to the organization, such as:

a) the monitoring and recording of process variables and product and service characteristics;

b) risk assessments of processes, products and services;

c) performance reviews, including on external providers and partners;

欠である．パフォーマンス指標を使用するステップを図4に示す．

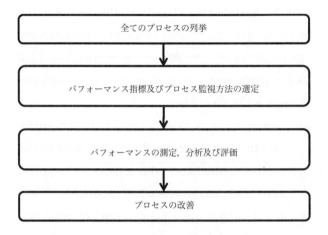

図4－パフォーマンス指標を使用するステップ

10.2.2 パフォーマンス指標に関する情報の収集に使用する方法は，次の例のように，組織にとって実用的及び適切であることが望ましい．

a) プロセス変数並びに製品及びサービスの特性の監視及び記録
b) プロセス，製品及びサービスに関するリスクアセスメント
c) 外部提供者及びパートナを含む，パフォーマンスのレビュー

d) interviews, questionnaires and surveys on the satisfaction of interested parties.

10.2.3 Factors that are within the control of the organization and critical to its sustained success should be subject to measurement and identified as key performance indicators (KPIs). These measurable KPIs should be:

a) accurate and reliable, to enable the organization to set measurable objectives, monitor and predict trends, and take actions for improvement and innovations when necessary;

b) selected as a basis for making strategic and operational decisions;

c) suitably cascaded as performance indicators at relevant functions and levels within the organization, to support the achievement of top level objectives;

d) appropriate to the nature and size of the organization, its products and services, processes and activities;

e) consistent with the strategy and objectives of the organization.

d) 利害関係者の満足度に関するインタビュー,アンケート及び調査

10.2.3 組織の管理下にあり,組織の持続的成功にとって必要不可欠な要因は,測定の対象とし,主要パフォーマンス指標(以下,KPI という.)として定義することが望ましい.これらの測定可能なKPI は,次のようなものであることが望ましい.

a) 組織が測定可能な目標を設定し,傾向を監視及び予測し,必要な場合には,改善及び革新への処置をとることができるほど,正確であり,信頼できる.

b) 戦略的及び運用上の決定を行うための基礎として選定している.

c) 最上位の目標の達成を支援するため,組織内の関連する部門及び階層において,パフォーマンス指標として適切に順次展開している.

d) 組織の性質及び規模,製品及びサービス,プロセス並びに活動に適している.

e) 組織の戦略及び目標と整合している.

10.2.4 The organization should consider specific information relating to risks and opportunities when selecting KPIs. In addition, the organization should ensure that KPIs provide information to implement action plans when performance does not achieve the objectives, or to improve and innovate process efficiency and effectiveness. Such information should take into account elements such as:

a) the needs and expectations of interested parties;
b) the importance of individual products and services to the organization;
c) the effectiveness and efficiency of processes;
d) the effective and efficient use of resources;
e) financial performance;
f) compliance with applicable external requirements.

10.3 Performance analysis

Analysis of the organization's performance should enable identification of issues, such as:

a) insufficient or ineffective resources within the organization;
b) insufficient or ineffective competencies, or-

10 組織のパフォーマンスの分析及び評価

10.2.4 組織は,KPIの選定に際して,リスク及び機会に関する固有の情報を考慮することが望ましい.さらに,組織は,パフォーマンスが目標を達成しない場合に,実施計画を行うための情報,又はプロセスの効率及び有効性を改善し,刷新するための情報を,KPIが提供することを確実にすることが望ましい.そのような情報には,次のような要素を考慮することが望ましい.

a) 利害関係者のニーズ及び期待

b) 個々の製品及びサービスの,組織にとっての重要性

c) プロセスの有効性及び効率

d) 資源の効果的及び効率的な利用

e) 財務パフォーマンス

f) 外部の適用可能な要求事項の遵守

10.3 パフォーマンス分析

組織のパフォーマンスの分析によって,次のような課題の特定が可能となることが望ましい.

a) 組織内での不十分又は非効果的な資源

b) 不十分若しくは非効果的な力量及び組織の知

ganizational knowledge and inappropriate behaviour;

c) risks and opportunities that are not being sufficiently addressed by the organization's management systems;

d) weakness in leadership activities, including:

   1) policy establishment and communication (see **Clause 7**);
   2) the managing of processes (see **Clause 8**);
   3) the managing of resources (see **Clause 9**);
   4) improvement, learning and innovation (see **Clause 11**);

e) potential strengths that might need to be fostered with regard to leadership activities;

f) processes and activities showing outstanding performance that could be used as a model to improve other processes.

The organization should have a clear framework to demonstrate the interrelations between its leadership activities and their effects on the organization's performance. This can enable the organization to analyse the strengths and weaknesses of its

識,並びに不適切な行動

c) 組織のマネジメントシステムによっては十分に取り組めていないリスク及び機会

d) 次の事項を含む,リーダーシップ活動における弱み
  1) 方針の策定及びコミュニケーション(箇条 **7** 参照)
  2) プロセスのマネジメント(箇条 **8** 参照)
  3) 資源のマネジメント(箇条 **9** 参照)
  4) 改善,学習及び革新(箇条 **11** 参照)

e) リーダーシップ活動に関して,伸ばす必要のありそうな潜在的な強み

f) 他のプロセスを改善するためのモデルとして使用することができる,傑出したパフォーマンスを示すプロセス及び活動

組織は,組織のリーダーシップ活動とそれらが組織のパフォーマンスに与える影響との相互関係を実証するための,明確な枠組みをもつことが望ましい.これによって,組織はそのリーダーシップ活動の強み・弱みを分析することができるようになる.

leadership activities.

10.4 Performance evaluation

10.4.1 The organization's performance should be evaluated from the viewpoint of the needs and expectations of its interested parties. When deviations from the needs and expectations are found, the processes and their interactions that affect its performance should be identified and analysed.

10.4.2 The organization's performance results should be evaluated against applicable objectives (see **7.3**) and their pre-determined criteria. Where objectives have not been achieved, the cause(s) should be investigated, with appropriate reviews of the deployment of the organization's policies, strategy and objectives and the organization's managing of resources, as necessary. Similarly, when objectives have been exceeded, what made it possible should be analysed in order to maintain the performance.

The results of evaluation should be understood by top management. Any identified performance fail-

10.4 パフォーマンス評価

10.4.1 組織のパフォーマンスは，利害関係者のニーズ及び期待という視点から評価することが望ましい．ニーズ及び期待からの逸脱が見つかった場合，パフォーマンスに影響を与えるプロセス及びその相互作用を特定し，分析することが望ましい．

10.4.2 組織のパフォーマンスの結果は，該当する目標（**7.3** 参照）及び目標について事前に決定された基準に照らして評価することが望ましい．目標が達成されていない場合には，その原因を調査し，必要に応じて，組織の方針，戦略及び目標の展開，並びに組織の資源のマネジメントについて，適切なレビューを行うことが望ましい．同様に，目標を超過している場合には，パフォーマンスを維持するため，それが可能になった要因を分析することが望ましい．

トップマネジメントは，評価の結果を理解することが望ましい．パフォーマンスについての特定され

ures should be prioritized for corrective action, based on the impact on the organization's policies, strategy and objectives.

Improvement achieved on the organization's performance should be evaluated from a long-term perspective. When the degree of improvement does not match the expected level, the organization should review the deployment of its policies, strategy and objectives for improvement and innovation, as well as the competencies and engagement of its people.

10.4.3 The organization's performance should be compared to established or agreed benchmarks. Benchmarking is a measurement and analysis methodology that an organization can use to search for the best practices inside and outside the organization, with the aim of improving its performance and innovative practices. Benchmarking can be applied to policies, strategy and objectives, processes and their operations, products and services, or the organization's structures.

たあらゆる未達成は，組織の方針，戦略及び目標に対する影響に基づき，是正処置のために優先付けすることが望ましい．

組織のパフォーマンスについて達成された改善を，長期的な展望から評価することが望ましい．改善の程度が期待されるレベルと合っていない場合には，組織は，改善及び革新に関する，組織の方針，戦略及び目標の展開，並びに人々の力量及び積極的参加について，レビューすることが望ましい．

10.4.3 組織のパフォーマンスを，確立した又は合意したベンチマークと比較することが望ましい．ベンチマーキングとは，組織が，そのパフォーマンスの改善及び革新的実践を目指して，組織内外のベストプラクティスを模索するために利用することができる測定及び分析の手法である．ベンチマーキングは，方針，戦略及び目標，プロセス及びその運用，製品及びサービス，又は組織構造に適用し得る．

10.4.4 The organization should establish and maintain a methodology for benchmarking that defines rules for items, such as:

a) the definition of the scope of the subject for benchmarking;

b) the process for choosing benchmarking partner(s), as well as any necessary communications and confidentiality policies;

c) the determination of indicators for the characteristics to be compared and the data collection methodology to be used;

d) the collection and analysis of data;

e) the identification of performance gaps and the indication of potential improvement areas;

f) the establishment and monitoring of corresponding improvement plans;

g) the inclusion of gathered experience into the organization's knowledge base and learning process (see **11.3**).

10.4.5 The organization should consider the different types of benchmarking practices, such as:

a) internal benchmarking of activities and processes within the organization;

10 組織のパフォーマンスの分析及び評価　127

10.4.4 組織は，次のような項目に関する取決めを定めたベンチマーキングの方法論を確立し，維持することが望ましい．

a) ベンチマーキングの適用範囲の定義

b) あらゆる必要なコミュニケーション及び機密保持に関する方針だけでなく，ベンチマーク先を選定するためのプロセス

c) 比較する特性に対する指標及び使用するデータの収集法の決定

d) データの収集及び分析
e) パフォーマンスのギャップの特定及び改善の可能性のある領域の提示
f) 対応する改善計画の策定及び監視

g) 蓄積された経験の組織の知識基盤及び学習プロセスへの取込み（**11.3** 参照）

10.4.5 組織は，次のような様々な種類のベンチマーキングの実践を検討することが望ましい．

a) 組織内での活動及びプロセスについての内部ベンチマーキング

b) competitive benchmarking of performance or processes with competitors;

c) generic benchmarking, by comparing strategies, operations or processes with unrelated organizations.

10.4.6 When establishing a benchmarking process, the organization should take into account that successful benchmarking depends on factors such as:

a) support from top management (as it involves mutual knowledge interchange between the organization and its benchmarking partners);

b) the methodology used to apply benchmarking;

c) an estimation of benefits versus costs;

d) an understanding of the characteristics of the subject being investigated, in order to allow a correct comparison with the current situation in the organization;

e) implementing lessons learned to bridge any determined gaps.

10.5 Internal audit

Internal audits are an effective tool for determining the levels of conformity of the organization's

10 組織のパフォーマンスの分析及び評価

b) 競合他社とのパフォーマンス又はプロセスについての競争的ベンチマーキング
c) 無関係な組織との戦略,運用又はプロセスの比較による,一般的なベンチマーキング

10.4.6 ベンチマーキングプロセスを確立する場合,組織は,ベンチマーキングの成功が次のような要因に依存している点を考慮することが望ましい.

a) トップマネジメントからの支援(組織とそのベンチマーク先との間の相互の知識交流を伴うため)
b) ベンチマーキングの適用に用いる方法論
c) 便益対コストの見積り
d) 組織の現状との正確な比較を可能にするための,調査対象の特性の理解

e) 明確にしたあらゆるギャップを埋めるための教訓の実施

10.5 内部監査

内部監査は,選定された基準に対する組織のマネジメントシステムの適合のレベルを明確にするため

management system to its selected criteria. They provide valuable information for understanding, analysing and improving the organization's performance. Internal audits should assess the implementation, effectiveness and efficiency of the organization's management systems. This can include auditing against more than one management system standard, as well as addressing specific requirements relating to interested parties, products, services, processes or specific issues.

To be effective, internal audits should be conducted in a consistent manner, by competent people, in accordance with the organization's audit planning. Audits should be conducted by people who are not involved in the activity being examined, in order to give an independent view on what is being performed.

Internal auditing is an effective tool for identifying problems, nonconformities, risks and opportunities, as well as for monitoring progress on resolving previously identified problems and nonconformities. Internal auditing can also be focused

の効果的なツールである．それによって，組織のパフォーマンスを理解し，分析し，改善するために貴重情報が得られる．内部監査は，組織のマネジメントシステムの実施，有効性及び効率を評価することが望ましい．これには，複数のマネジメントシステム規格に対する監査，及び利害関係者，製品，サービス，プロセス又は特定の課題に関連する固有の要求事項を取り扱う監査を含むことができる．

効果的な内部監査のために，内部監査は，組織の監査計画に従って力量のある人々が，整合性のある方法で実施することが望ましい．監査は，実施していることに対して独立性をもった視点を与えるために，評価の対象となっている活動に関与していない人々が実施することが望ましい．

内部監査は，以前に特定された問題及び不適合の解決に関する進捗状況を監視するだけでなく，問題，不適合，リスク及び機会を特定するための効果的なツールである．また，内部監査は，優れた実践の特定及び改善の機会に焦点を合わせることもでき

on the identification of good practices and on improvement opportunities.

The outputs of internal audits provide a useful source of information for:
a) addressing problems, nonconformities and risks;
b) identifying opportunities;
c) promoting good practices within the organization;
d) increasing understanding of the interactions between processes.

Internal audit reporting usually contains information on conformity to the given criteria, nonconformities and improvement opportunities. Audit reporting is an essential input for management review. Top management should establish a process for reviewing all internal audit results, in order to identify trends that can require organization-wide corrective actions and opportunities for improvement.

The organization should use the results of other audits, such as second- and third-party audits,

10 組織のパフォーマンスの分析及び評価　133

る．

　内部監査のアウトプットは，次の事項に役立つ情報源を提供する．
a) 問題，不適合及びリスクへの取組み
b) 機会の特定
c) 組織内の優れた実践の普及

d) プロセス間の相互作用に関する理解の向上

　内部監査の報告は，通常，与えられた基準への適合，不適合及び改善の機会に関する情報を含む．また，監査報告書は，マネジメントレビューへの必要不可欠なインプットである．トップマネジメントは，組織全体にわたる是正処置を必要とするような傾向及び改善の機会を特定するために，全ての内部監査の結果をレビューするプロセスを確立することが望ましい．

　組織は，是正処置のためのフィードバックとして，第二者監査及び第三者監査のような他の監査の

as feedback for corrective actions. It can also use them to monitor progress in the implementation of appropriate plans intended to facilitate the resolution of nonconformities, or for the implementation of identified opportunities for improvement.

NOTE   See ISO 19011 for further guidance on auditing management systems.

10.6 Self-assessment

Self-assessment should be used to determine the strengths and weaknesses of the organization as well as best practices, both at an overall level and at the level of individual processes. Self-assessment can assist the organization to prioritize, plan and implement improvements and/or innovations, where necessary.

Elements of a management system should not be assessed independently given that processes are interdependent. This allows for assessment of relationships between elements and their impacts on the organization's mission, vision, values and culture.

結果を活用することが望ましい．また，そうした監査結果を，不適合の解決を容易にすることを意図した，又は特定された改善の機会を実施するための，適切な計画の実施における進捗状況を監視するために，活用することもできる．

> **注記** マネジメントシステム監査に関する追加の手引については，**JIS Q 19011**を参照．

10.6 自己評価

自己評価は，組織全体のレベル及び個々のプロセスレベルの両方における，組織のパフォーマンスの強み・弱み及びベストプラクティスを明確にするために利用することが望ましい．自己評価は，必要に応じて，組織が，改善及び／又は革新の優先順位を付け，計画し，実施することの手助けとなり得る．

プロセスが相互依存している場合には，マネジメントシステムの要素を独立して評価しないほうがよい．これによって，要素と，要素が組織の使命，ビジョン，価値観及び文化に与える影響との関係を評価できるようになる．

The results of self-assessment support:

a) improvement of the organization's overall performance;

b) progress towards achieving and maintaining sustained success for the organization;

c) innovation in the organization's processes, products and services, and the organization's structure, when appropriate;

d) recognition of best practices;

e) identification of further opportunities for improvement.

The results of self-assessment should be communicated to relevant people within the organization, in order to be used to share understanding about the organization and its future direction.

A self-assessment tool based on this document is provided in **Annex A**.

10.7 Reviews

Reviews of performance measurement, benchmarking, analysis and evaluations, internal audits and self-assessments should be performed by ap-

自己評価の結果は，次の事項を支援する．

a) 組織の全体的なパフォーマンスの改善

b) 組織の持続的成功の達成及び維持に向けた進展

c) 必要に応じた，組織のプロセス，製品及びサービス並びに組織構造の革新

d) ベストプラクティスの認知

e) 改善のための更なる機会の特定

自己評価の結果は，組織及びその今後の方向性についての理解を共有するのに利用するため，組織内の関連する人々に伝達することが望ましい．

この規格に基づいた自己評価ツールを，**附属書 A** に記載する．

10.7 レビュー

パフォーマンスの測定，ベンチマーキング，分析及び評価，内部監査並びに自己評価についてのレビューを，組織の適切な階層及び部門，並びにトップ

propriate levels and functions of the organization, as well as by top management. The reviews should be conducted at planned and periodic intervals, to enable trends to be determined and to evaluate the organization's progress towards achieving its policies, strategy and objectives. They should also address the assessment and evaluation of improvement, learning and innovation activities performed previously, including aspects of adaptability, flexibility and responsiveness in relation to the organization's mission, vision, values and culture.

The reviews should be used by the organization to understand the needs of adapting its policies, strategy and objectives (see **Clause 7**). They should also be used to determine the opportunities for improvement, learning and innovation of the organization's managerial activities (see **Clause 11**).

The reviews should enable evidence-based decision making and the establishment of actions to achieve desired results.

11 Improvement, learning and innovation

マネジメントが実施することが望ましい．レビューは，その傾向を明確にできるよう，また，組織の方針，戦略及び目標の達成へ向けた進捗状況を評価するために，あらかじめ定められた，定期的な間隔で実施することが望ましい．レビューにおいては，組織の使命，ビジョン，価値観及び文化との関連における適応性，柔軟性及び応答性の側面を含め，それまでに実施した改善，学習，革新活動の診断及び評価に取り組むことが望ましい．

組織は，その方針，戦略及び目標を適応させる必要性を理解するため，レビューを利用することが望ましい（箇条7参照）．また，レビューは，組織のマネジメント活動の改善，学習及び革新の機会を明確にするために利用することが望ましい（箇条11参照）．

レビューによって，証拠に基づく意思決定及び望ましい結果を達成するための処置の策定を可能にすることが望ましい．

11 改善，学習及び革新

11.1 General

Improvement, learning and innovation are interdependent and key aspects that contribute to the sustained success of an organization. They create inputs into products, services, processes and management systems, and contribute to achieving desired results.

The organization will experience constant change in its external and internal issues and in the needs and expectations of its interested parties. Improvement, learning and innovation support the organization's ability to respond to these changes in a manner that enables it to fulfil its mission and vision, as well as supporting its achievement of sustained success.

11.2 Improvement

Improvement is an activity to enhance performance. Performance can relate to a product or service, or to a process. Improving product or service performance or the management system can help the organization anticipate and meet the needs and expectations of interested parties and also in-

11.1 一般

改善,学習及び革新は,相互に依存しており,組織の持続的成功に貢献する重要な側面である.これらは,製品,サービス,プロセス及びマネジメントシステムへのインプットを生み出し,望ましい結果の達成に貢献する.

組織は,その外部及び内部の課題,並びに利害関係者のニーズ及び期待における変化を絶えず受ける.改善,学習及び革新は,持続的成功の達成を支援するだけでなく,組織がその使命及びビジョンを満たすことができるよう,こうした変化に対応する組織の能力を支援する.

11.2 改善

改善とは,パフォーマンスを向上させる活動である.パフォーマンスは,製品若しくはサービス,又はプロセスと関係し得る.製品若しくはサービスのパフォーマンス又はマネジメントシステムの改善は,組織が,利害関係者のニーズ及び期待を予想し,満たし,経済的効率を上げるために役立ち得

crease economic efficiency. Improving processes can lead to increased effectiveness and efficiency, resulting in benefits such as cost, time and energy saving and reduced waste; in turn, this can lead to meeting the needs and expectations of interested parties more effectively.

Improvement activities can range from small-step continual improvements to significant improvements of the entire organization.

The organization should define objectives for improving its products or services, processes, structure and management system, by using the results of the analysis and evaluation of its performance.

Improvement processes should follow a structured approach. The methodology should be applied consistently for all processes.

The organization should ensure that improvement becomes established as a part of the organization's culture by:

a) empowering people to participate in and con-

11 改善,学習及び革新

る.プロセスの改善は,有効性及び効率の増加につながり,結果として,コスト,時間,エネルギー及び無駄の削減などの便益をもたらし,更には,利害関係者のニーズ及び期待をより効果的に満たすことにつながり得る.

改善活動は,小さな継続的改善から組織全体の大きな改善まで広範囲にわたり得る.

組織は,そのパフォーマンスの分析及び評価の結果を活用して,その製品又はサービス,プロセス,構造及びマネジメントシステムの改善目標を定めることが望ましい.

改善プロセスは,構造化されたアプローチに従うことが望ましい.この方法論は,全てのプロセスに対して整合して適用することが望ましい.

組織は,改善が,次の事項によって,組織文化の一部として確立されるようになることを確実にすることが望ましい.
a) 人々が改善の取組みに参加し,その達成の成功

tribute to the successful achievement of improvement initiatives;

b) providing the necessary resources to achieve improvements;

c) establishing recognition systems for improvements;

d) establishing recognition systems for improving the effectiveness and efficiency of the improvement process;

e) engagement of top management in improvement activities.

11.3 Learning

11.3.1 The organization should encourage improvement and innovation through learning. The inputs for learning can be derived from many sources, including experience, analysis of information, and the results of improvements and innovations.

A learning approach should be adopted by the organization as a whole, as well as at a level that integrates the capabilities of individuals with those of the organization.

に貢献するための権限委譲

b) 改善を達成するために必要な資源の提供

c) 改善に対する表彰制度の確立

d) 改善プロセスの有効性及び効率を改善するための表彰制度の確立

e) 改善活動へのトップマネジメントの積極的参加

11.3 学習
11.3.1 組織は,学習を通した,改善及び革新を奨励することが望ましい.学習へのインプットは,経験,情報の分析,並びに改善及び革新の結果を含む多くの情報源から得られる.

組織は,学習アプローチを,個人の実現能力を組織の実現能力へ統合したレベルで採用するだけでなく,組織全体として採用することが望ましい.

11.3.2 Learning as an organization involves consideration of:

a) collected information relating to various external and internal issues and interested parties, including success stories and failures;

b) insight through in-depth analysis of the information collected.

Learning that integrates the capabilities of individuals with those of the organization is achieved by combining the knowledge, thinking patterns and behaviour patterns of people with the values of the organization.

Knowledge can be explicit or tacit. It can originate from inside or outside the organization. It should be managed and maintained as an asset of the organization.

The organization should monitor its organizational knowledge and determine the need to acquire, or more effectively share, knowledge throughout the organization.

11 改善,学習及び革新

11.3.2 組織としての学習は,次の事項を考慮することを含む.

a) 成功事例及び失敗事例を含む,様々な外部及び内部の課題並びに利害関係者に関連する,収集した情報

b) 収集した情報の徹底的な分析から得られた洞察

　個人の実現能力を組織の実現能力へ統合する学習は,人々の知識,思考パターン及び行動パターンと組織の価値観とを組み合わせることによって達成される.

　知識には,形式知又は暗黙知があり得る.知識は,組織内外から生じる可能性がある.知識は,組織の資産としてマネジメントし,維持することが望ましい.

　組織は,その組織の知識を監視し,組織全体を通じて,知識を獲得する,又はより効果的に共有する必要性を明確にすることが望ましい.

11.3.3 In order to foster a learning organization, the following factors should be considered:

a) the organization's culture, aligned with its mission, vision and values;

b) top management supporting initiatives in learning, by demonstrating its leadership and through its behaviour;

c) stimulation of networking, connectivity, interactivity and sharing of knowledge both inside and outside the organization;

d) maintaining systems for learning and sharing of knowledge;

e) recognizing, supporting and rewarding the improvement of people's competence, through processes for learning and sharing of knowledge;

f) appreciating creativity and supporting diversity of the opinions of the different people in the organization.

Rapid access to, and use of, organizational knowledge can enhance the organization's ability to manage and maintain its sustained success (see **9.3**).

11 改善,学習及び革新

11.3.3 学習する組織を育成するため,次の要因を考慮することが望ましい.

a) 組織の使命,ビジョン及び価値観と一貫性のある組織文化

b) トップマネジメントがそのリーダーシップを発揮することによって,及びその行動を通じて,学習への取組みを支援すること

c) 組織の内外におけるネットワーク作り,人々のつながり,相互作用及び知識の共有の促進

d) 学習及び知識の共有のためのシステムの維持

e) 学習及び知識共有のためのプロセスを通じて,人々の力量の改善を認め,支援し,褒賞を与えること

f) 創造性を認め,組織における異なる人々の意見の多様性を支援すること

組織の知識に迅速にアクセスし,利用することは,組織がその持続的成功をマネジメントし,維持する能力を高めることができる(**9.3**参照).

11.4 Innovation

11.4.1 General

Innovation should result in improvement leading to new or changed products or services, processes, market position, or performance, enabling realization or redistribution of value.

Changes in the organization's external and internal issues and the needs and expectations of interested parties could require innovation.

To support and promote innovation the organization should:

a) identify specific needs for innovation and encourage innovative thinking in general;
b) establish and maintain processes that allow for effective innovation;
c) provide the resources needed to realize innovative ideas.

11.4.2 Application

Innovation can be applied at all levels of the organization, through changes in:

a) technology or products or services (i.e. inno-

11.4 革新

11.4.1 一般

革新は，価値の実現又は再配布を可能にする，新規又は変更された製品若しくはサービス，プロセス，市場における位置付け，又はパフォーマンスにつながる改善をもたらすことが望ましい．

組織の外部及び内部の課題，並びに利害関係者のニーズ及び期待における変化によって，革新を必要とすることがある．

組織は，革新を支援し，促進するために，次の事項を行うことが望ましい．

a) 革新に対する固有のニーズを特定し，全般的な革新的思考を奨励する．
b) 効果的な革新を可能にするプロセスを確立し，維持する．
c) 革新的なアイデアを実現するために必要な資源を提供する．

11.4.2 適用

革新は，次の事項の変化に応じて，組織の全ての階層で適用することができる．

a) 技術，又は製品若しくはサービス（すなわち，

vations that not only respond to the changing needs and expectations of interested parties, but also anticipate potential changes in the organization and in the life cycles of its products and services);

b) processes (i.e. innovation in the methods for production and service provision, or innovation to improve process stability and reduce variation);

c) the organization (i.e. innovation in the constitution and the structures of the organization);

d) the organization's management system (i.e. to ensure that competitive advantage is maintained and new opportunities are utilized, when there are emerging changes in the organization's context);

e) the organization's business model (i.e. innovation in responding to distribution of value to customers or changing market position in accordance with interested parties' needs and expectations).

11.4.3 Timing and risk

The organization should evaluate the risks and op-

利害関係者の変化するニーズ及び期待に応えるだけでなく，組織及びその製品又はサービスのライフサイクルに起こり得る変化を先取りする革新）

b) プロセス（すなわち，製造及びサービス提供の方法における革新，又はプロセスの安定度を改善し，ばらつきを減少させる革新）

c) 組織（すなわち，組織体質及び組織構造の革新）

d) 組織のマネジメントシステム（すなわち，組織の状況に変化が起こっている場合に，競争優位を維持し，新たな機会を活用することを確実にするための革新）

e) 組織のビジネスモデル（すなわち，利害関係者のニーズ及び期待に従った，顧客への価値の分配又は変化する市況への対応における革新）

11.4.3 タイミング及びリスク

組織は，革新活動のための計画に関連するリスク

portunities related to its plans for innovation activities. It should give consideration to the potential impact on the managing of changes and prepare action plans to mitigate those risks (including contingency plans), where necessary.

The timing for the introduction of an innovation should be aligned with the evaluation of the risk associated with undertaking that innovation. The timing should usually be a balance between the urgency with which it is needed and the resources that are made available for its development.

The organization should review, improve and innovate based on the results of its performance evaluation (see **Clause 10**).

The organization should use a process that is aligned with its strategic direction to plan and prioritize innovation initiatives.

The results of innovation should be reviewed in order to experience learning and to increase organizational knowledge.

及び機会を評価することが望ましい．組織は，変更のマネジメントに対する潜在的な影響を考慮し，必要な場合には，（不測の事態に対する対応を含む）そうしたリスクを軽減するための実施計画を準備することが望ましい．

革新を行うタイミングは，その革新の実施に関連するリスクの評価と一貫させることが望ましい．そのタイミングは，通常，革新が必要とされる緊急性と，革新の展開のために利用可能とすべき資源とのバランスが保たれていることが望ましい．

組織は，そのパフォーマンス評価の結果に基づいてレビューし，改善し，革新することが望ましい（箇条 **10** 参照）．

組織は，その戦略的方向性と一貫しているプロセスを用いて革新への取組みを計画し，その優先順位を付けることが望ましい．

学習を経験し，組織の知識を増加させるため，革新の結果をレビューすることが望ましい．

注記 改善に関するより詳細な指針を定めた規格として,**JIS Q 9024** がある.

Annex A

(informative)

Self-assessment tool

A.1 General

Self-assessment can provide an overall view of the performance of an organization and the degree of maturity of its management system. It can help to identify areas for improvement and/or innovation and to determine priorities for subsequent actions.

Audits are used to determine the extent to which, for example, requirements related to a management system are fulfilled (against a defined standard, or the organization's own criteria). Audit findings are used to assess the effectiveness of, for example, a management system, and to identify risks and opportunities for improvement.

An organization should use self-assessment to identify improvement and innovation opportunities, set priorities and establish action plans, with the objective of sustained success. The output of

附属書 A
（参考）
自己評価ツール

A.1 一般

自己評価は，組織のパフォーマンス及びマネジメントシステムの成熟度について，全体像を提供することができる．自己評価は，改善及び／又は革新を必要とする領域を特定すること，並びにそれに続く行動の優先順位の決定に役立てることができる．

一方，監査は，例えば，マネジメントシステムに関連する要求事項が満たされている程度を明確にするために使用される（定められている標準又は組織自身の基準に照らして）．監査所見は，例えば，マネジメントシステムの有効性を評価し，リスク及び改善の機会を特定するために使用される．

組織は，改善及び革新の機会を特定し，優先順位を付け，持続的成功の目標に伴う実施計画を策定するために，自己評価を利用することが望ましい．自己評価のアウトプットは，組織の強み・弱み，関係

self-assessment will show strengths and weaknesses, the related risks and opportunities for improvement, the maturity level of the organization and, if repeated, the organization's progress over time.

The results of an organization's self-assessment can be a valuable input into management reviews. Self-assessment also has the potential to be a learning tool, which can provide an improved overview of the organization, promote the involvement of interested parties and support the overall planning activities of the organization.

The self-assessment tool given in this annex is based on the guidance detailed in this document and provides a framework for improvement. It can be used as given, or it can be customized to suit the organization.

A.2 Maturity model

A mature organization performs effectively and efficiently and achieves sustained success by:

a) understanding and satisfying the needs and

する改善のためのリスク及び機会,組織の成熟度レベル,及び自己評価を繰り返す場合には,長期にわたる組織の進捗状況を示す.

組織の自己評価の結果は,マネジメントレビューへの価値あるインプットとなり得る.また,自己評価は,組織の改善された全体像を提供し,利害関係者の参画を促進し,組織の全体的な計画活動を支援することができる学習ツールになる可能性をもっている.

この附属書に記載している自己評価ツールは,この規格に記載している手引に基づくもので,改善の枠組みを与える.このツールは,記載しているとおり使用すること,又は組織に合うように調整することが可能である.

A.2 成熟度モデル

成熟した組織は,次の事項によって効果的及び効率的に成果を挙げ,持続的成功を達成する.

a) 利害関係者のニーズ及び期待を理解し,これら

expectations of interested parties;

b) monitoring changes in the context of the organization;
c) identifying possible areas for improvement, learning and innovation;
d) defining and deploying policies, strategy and objectives;
e) managing its processes and resources;
f) demonstrating confidence in its people, leading to increased engagement;
g) establishing beneficial relationships with interested parties, such as external providers and other partners.

This self-assessment tool uses five maturity levels, which can be extended to include additional levels or otherwise customized as needed.

Table A.1 gives a generic framework for setting out how performance criteria can be related to the levels of maturity in a tabular format. The organization should review its performance against the specified criteria, identify its current maturity levels, and determine its strengths and weakness-

附属書A（参考）　　　　163

を満たす．
b) 組織の状況における変化を監視する．

c) 改善，学習及び革新が可能な領域を特定する．

d) 方針，戦略及び目標を決定し，展開する．

e) 組織のプロセス及び資源をマネジメントする．
f) 人々に対する信頼を示し，積極的参加の向上につなげる．
g) 外部提供者及びその他のパートナなどの利害関係者と有益な関係を確立する．

　この自己評価ツールでは，5段階の成熟度レベルを用いており，追加レベルを含める拡張を行うか，又は必要に応じて調整することが可能である．

　表形式を用いてどのようにパフォーマンス基準と成熟度レベルとを関連付けることができるかを表す一般的枠組みを**表A.1**に示す．組織は，特定の基準に照らしてそのパフォーマンスをレビューし，組織の現在の成熟度を特定し，その強み・弱み，並びに関連するリスク及び機会の改善を明確にすること

es and the related risks and opportunities for improvement.

The criteria given for the higher levels can help the organization to understand the issues it needs to consider and to determine the improvements needed to reach higher levels of maturity. **Tables A.2** to **A.32** give self-assessment criteria based on this document.

Table A.1 — **Generic model for self-assessment elements and criteria related to maturity levels**

Maturity level towards sustained success					
Key element	Level 1	Level 2	Level 3	Level 4	Level 5
Element 1	Criteria 1 Base level				Criteria 1 Best practice
Element 2	Criteria 2 Base level				Criteria 2 Best practice
Element 3	Criteria 3 Base level				Criteria 3 Best practice

A.3 Self-assessment of detailed elements

This self-assessment is intended to be performed by process owners and managers at all levels to obtain an in-depth overview of the organization and

が望ましい.

より高いレベルとして与えられた基準は,検討を要する課題を理解し,組織がより高いレベルの成熟度に達する上で必要となる改善を明確にするために役立ち得る.**表A.2〜表A.32**は,この規格に基づいた自己評価基準を示す.

表A.1－成熟度レベルに関連する自己評価の要素及び基準の一般モデル

持続的成功に至る成熟度レベル					
主要要素	レベル1	レベル2	レベル3	レベル4	レベル5
要素1	基準1 基本レベル				基準1 ベストプラクティス
要素2	基準2 基本レベル				基準2 ベストプラクティス
要素3	基準3 基本レベル				基準3 ベストプラクティス

A.3 各箇条に対する自己評価

この自己評価は,組織及び組織の現在のパフォーマンスの全体像を詳細に把握するため,プロセスオーナ及び全ての階層の管理者が実施することを意図

its current performance.

The elements of this self-assessment are contained in **Tables A.2** to **A.32** and relate to the subclauses of this document; however, the organization can define additional or different criteria to fulfil its own specific needs. If appropriate, the self-assessment can be limited to any of the tables in isolation.

A.4 Using the self-assessment tools

A.4.1 The purposes of a step-by-step methodology for an organization to conduct a self-assessment are:

a) to define the scope of the self-assessment in terms of the parts of the organization to be assessed and the type of the assessment, such as:

   1) a self-assessment of key elements;
   2) a self-assessment of detailed elements, based on this document;
   3) a self-assessment of detailed elements, based on this document with additional or new criteria or levels;

している.

この自己評価の要素は,表 A.2 〜表 A.32 に記載しており,この規格の本体の各細分箇条に対応している.しかしながら,組織は,その固有のニーズを満たすために,付加的又は個別の基準を定めることができる.適切な場合,表 A.2 〜表 A.32 の必要な表に限定して自己評価することができる.

A.4 自己評価ツールの活用

A.4.1 組織が自己評価を実施するための逐次的な方法は,次のとおりである.

a) 自己評価の範囲,すなわち,組織の部門及び次のような評価の種類を定める.

1) 主要要素の自己評価
2) この規格に基づく各箇条・細分箇条に対する自己評価
3) 付加的又は個別の基準又はレベルを取り入れた,この規格に基づく,各箇条・細分箇条に対する自己評価

b) to identify who will be responsible for the self-assessment and when it will be carried out;

c) to determine how the self-assessment will be carried out, either by a team (cross-functional or other appropriate team) or by individuals (the appointment of a facilitator can assist the process);

d) to identify the maturity level for each of the organization's individual processes, which should be done by:

   1) comparing the present situation in the organization to the scenarios that are listed in the tables;

   2) marking the elements that the organization is already applying, i.e. start at level 1 and build on progress, to attain maturity level 5 by incorporating the criteria identified in levels 3 and 4;

   3) establishing the current maturity level;

e) to consolidate the results into a report, which this provides a record of progress over time and can facilitate the communication of information, both externally and internally (the use of graphics in such a report can aid the

b) 誰が自己評価に責任をもつのか，及びいつ自己評価を実施するのかを決定する．

c) 自己評価をどのように実施するのか，チーム（部門横断又はその他の適切なチーム）によるのか，個人によるのかを決定する（自己評価の支援者を任命することによって，このプロセスを支援することができる．）．

d) 組織の個々のプロセスの成熟度レベルを特定する．これを，次の事項によって行うことが望ましい．

　1) 組織の現在の状況と表に記載したシナリオとの比較

　2) 組織が既に実施している要素に印を付けること．すなわち，レベル1から始め，レベル3及びレベル4で特定した基準を取り入れることによって，成熟度レベル5を達成するよう進歩を積み重ねる．

　3) 現在の成熟度レベルの確立

e) 結果を報告書にまとめる．これは，長期にわたる進歩の記録となり，組織内外の情報交換に役立たせることができる（このような報告書にグラフを使用することは，結果の伝達に有用である．）．

communication of the results);

f) to assess the current performance of the organization's processes and identify areas for improvement and/or innovation (these opportunities should be identified through the self-assessment process and an action plan developed as a result of the assessment).

A.4.2 An organization can be at different maturity levels for the different elements. A review of the gaps can help top management in planning and prioritizing the improvement and/or innovation activities needed to move individual elements to a higher level.

The completion of a self-assessment should result in an action plan for improvement and/or innovation that should be used as an input to top management for planning and review, based on the elements of this document.

The information gained from the self-assessment can also be used:

a) to stimulate comparisons and share learning

附属書 A（参考）

f) 組織のプロセスの現在のパフォーマンスを評価し，改善及び／又は革新すべき領域を特定する（これらの機会は，自己評価プロセス及び評価の結果によって策定する実施計画を通して特定することが望ましい．）．

A.4.2 組織の成熟度レベルは，要素ごとに異なり得る．レベルを規定している要素と組織の現状とのギャップのレビューは，トップマネジメントが個々の要素をより高いレベルに向上させるために必要な改善及び／又は革新活動を計画し，優先順位付けをすることに役立てることができる．

自己評価の完了は，この規格の要素に基づいて，改善及び／又は革新のための実施計画の策定につながり，それがトップマネジメントによる計画の策定及びレビューへのインプットとして利用することが望ましい．

また，自己評価から得られた情報は，次の事項にも利用できる．
a) 組織全体を通して，互いに比較し，組織全体に

throughout the organization (the comparisons can be between the organization's processes and, where applicable, between its different units);
b) to benchmark with other organizations;
c) to monitor progress of the organization over time, by conducting periodic self-assessments.

Following the reviews, the organization should assign responsibilities for the chosen actions, estimate and provide the resources needed, and determine the expected benefits and any perceived risks associated with them.

わたり学習を共有する(比較は,組織のプロセス間,及び該当する場合には,必ず,異なる事業単位間で行うことができる.).

b) 他の組織とのベンチマーキング
c) 定期的な自己評価を実施することによって,長期にわたる組織の進捗状況を監視する.

このレビューに従い,組織は,選択した行動に対する責任を割り当て,必要な資源の見積り及び提供を行い,期待する便益及びそれに付随する認識したリスク全てを明確にすることが望ましい.

Table A.2 — Self-assessment of the detailed elements of 5.2

Subclause	Maturity level		Conclusion	
	Level	Item[a]	YES	Results/comment[b]
5.2 Relevant interested parties	1	The interested parties are determined, including their needs and expectations and whether the associated risks and opportunities are informal or ad hoc.		
	2	Processes to meet the needs of some interested parties are established.		
		Ongoing relationships with interested parties are established as informal or ad hoc.		
	3	Processes for determining which interested parties are relevant are in place.		
		Processes for determining the relevance of interested parties include consideration of those that are a risk to sustained success if their needs and expectations are not met and those that can provide opportunities to enhance sustained success.		
		The needs and expectations of the relevant interested parties are identified.		
		Processes to fulfil the needs and expectations of the interested parties are established.		
	4	Processes for assessing the relevance of the needs and expectations for relevant interested parties are in place and are used to determine which ones need to be addressed.		

附属書A（参考）

表A.2 − 5.2に対する自己評価

細分箇条	成熟度レベル		結論	
	レベル	項目 a)	YES	結果／コメント b)
5.2 密接に関連する利害関係者	1	利害関係者をそのニーズ及び期待，また，関連するリスク及び機会を含め，非公式又はその場限りで決めている．		
	2	利害関係者のニーズを満たすための幾つかのプロセスを確立している．		
		非公式又はその場限りとして利害関係者との継続的な関係を確立している．		
	3	どの利害関係者が密接に関連するかを明確にするプロセスが整備されている．		
		利害関係者の関連性を明確にするプロセスは，利害関係者のニーズ及び期待を満たさない場合，持続的成功にとってリスクとなる人々及び持続的成功を高めるための機会を提供することができる人々についての考慮を含んでいる．		
		密接に関連する利害関係者のニーズ及び期待を特定している．		
		利害関係者のニーズ及び期待を満たすためのプロセスを確立している．		
	4	密接に関連する利害関係者にとってのニーズ及び期待の関係性を評価するプロセスを整備しており，どれに取り組む必要があるかを明確にするために利用している．		

Table A.2 *(continued)*

Subclause	Maturity level		Conclusion	
	Level	Item[a]	YES	Results/comment[b]
		The needs and expectations of key interested parties are addressed and reviewed such that improved performance, common understanding of objectives and values, and enhanced stability have been realized in some of these on-going relationships.		
	5	Processes and relationships with relevant interested parties are fulfilled according to the relevant needs and expectations determined. This has been done as part of understanding the benefits, risks and opportunities of ongoing relationships.		
		The needs and expectations of all relevant interested parties are addressed, analysed, evaluated and reviewed, such that there is improved and sustained performance, common understanding of objectives and values, and enhanced stability, including recognition of the benefits derived from these ongoing relationships.		

[a] Items outlined in levels 3 to 5 are intended to be a progression of thought that is based on the guidance provided in the applicable subclause.
[b] This may include recognition of aspects where the organization is partially meeting a maturity level.

附属書A（参考）

表 A.2（続き）

細分箇条	成熟度レベル		結論	
	レベル	項目 a)	YES	結果／コメント b)
		パフォーマンス改善，目標及び価値観の共通理解並びに安定性向上が，これらの継続的な関係の一部で実現するよう，重要な利害関係者のニーズ及び期待に取り組み，レビューしている．		
	5	プロセス及び密接に関連する利害関係者との関係を，明確にした関連するニーズ及び期待に従って遂行している．これは，継続的な関係の便益，リスク及び機会を理解することの一環として行われている．		
		全ての密接に関連する利害関係者のニーズ及び期待に取り組み，分析し，評価し，レビューしており，それによって，これらの継続的な関係から得られる便益の認識を含め，パフォーマンスの改善及び持続，目標及び価値観の共通理解並びに安定性向上を得ている．		

注 a) レベル3からレベル5までに記載している項目は，当該細分箇条に示した手引に基づいており，成熟度のレベルを示すことを意図している．
b) 結果／コメント欄は，組織が一つの成熟度レベルを部分的に満たしているという状況の認識を含んでもよい．

Table A.3 — Self-assessment of the detailed elements of 5.3

Subclause	Maturity level		Conclusion	
	Level	Item[a]	YES	Results/ comment[b]
5.3 External and internal issues	1	Processes for determining and addressing external and internal issues are informal or ad hoc.		
	2	Processes for determining and addressing issues are in place.		
		The risks and opportunities related to the issues identified are determined as informal or ad hoc.		
	3	Processes to determine internal issues that can affect the organization's ability to achieve sustained success are identified.		
		Processes to determine external issues that can affect the organization's ability to achieve sustained success are identified.		
	4	External and internal issues are determined and show consideration for factors such as statutory, regulatory and sector specific requirements, globalization, innovation, activities and associated processes, strategy and levels of competence and organizational knowledge.		
		Risks and opportunities are determined, and show consideration for information from the organization's past and its current situation.		

附属書 A (参考)

表 A.3 − 5.3 に対する自己評価

細分箇条	成熟度レベル		結論	
	レベル	項目 a)	YES	結果／コメント b)
5.3 外部及び内部の課題	1	外部及び内部の課題を明確にして取り組むプロセスは，非公式又はその場限りのものである．		
	2	課題を明確にして取り組むためのプロセスを整備している．		
		特定した課題に関わるリスク及び機会を，非公式又はその場限りのものとして明確にしている．		
	3	持続的成功を達成する組織の能力に影響を及ぼし得る内部の課題を明確にするためのプロセスを特定している．		
		持続的成功を達成する組織の能力に影響を及ぼし得る外部の課題を明確にするためのプロセスを特定している．		
	4	外部及び内部の課題を明確にしており，法令・規制要求事項及び分野固有の要求事項，グローバル化，革新，活動及び関連するプロセス，戦略並びに力量及び組織の知識のレベルのような要因に対する考慮を示している．		
		リスク及び機会を明確にしており，組織の過去からの情報及び現在の状況に対する考慮を示している．		

Table A.3 *(continued)*

Subclause	Maturity level		Conclusion	
	Level	Item[a]	YES	Results/comment[b]
		Processes to address issues considered to be risks to sustained success, or opportunities to enhance sustained success, are established, implemented and maintained.		
	5	Processes for the ongoing monitoring, reviewing and evaluation of external and internal issues are established, implemented and maintained, with actions arising from this process acted on.		

[a] Items outlined in levels 3 to 5 are intended to be a progression of thought that is based on the guidance provided in the applicable subclause.

[b] This may include recognition of aspects where the organization is partially meeting a maturity level.

附属書A（参考）

表 A.3（続き）

細分箇条	成熟度レベル		結論	
	レベル	項目 a)	YES	結果／コメント b)
		持続的成功へのリスク又は持続的成功を強化する機会を考慮した課題に取り組むためのプロセスを確立し，実施し，維持している．		
	5	外部及び内部の課題を継続的に監視し，レビューし，評価するプロセスを確立し，実施し，維持しており，このプロセスから生じる処置をとっている．		

注 a) レベル3からレベル5までに記載している項目は，当該細分箇条に示した手引に基づいており，成熟度のレベルを示すことを意図している．
 b) 結果／コメント欄は，組織が一つの成熟度レベルを部分的に満たしているという状況の認識を含んでもよい．

Table A.4 — Self-assessment of the detailed elements of 6.2

Subclause	Maturity level		Conclusion	
	Level	Item[a]	YES	Results/ comment[b]
6.2 Mission, vision, values and culture	1	Processes for determining the identity of the organization, along with the establishment of mission, vision, values and culture, are informal or ad hoc.		
	2	A basic understanding of the organization's mission, vision and values exists.		
		An understanding of the current culture, and whether there is a need to change it, is informal or ad hoc.		
	3	Top management is involved in determining the mission, vision and values, based on processes that account for the definition and sustainment of the context of the organization in relation to its defined identity.		
		An understanding of the current culture, along with a process for considering the need for a change to that culture, is in place.		
		Changes to the organization's identity are communicated informally to perceived interested parties.		
	4	The organization's culture is aligned with its mission, vision and values.		

表 A.4 – 6.2 に対する自己評価

細分箇条	成熟度レベル		結論	
	レベル	項目 a)	YES	結果／コメント b)
6.2 使命，ビジョン，価値観及び文化	1	使命，ビジョン，価値観及び文化の確立とともに，組織のアイデンティティを明確にするプロセスは，非公式又はその場限りである．		
	2	組織の使命，ビジョン及び価値観の基本的な理解がある．		
		現在の文化の理解及び文化を変化させる必要があるかどうかは，非公式又はその場限りである．		
	3	トップマネジメントは，定めたアイデンティティに関連する組織の状況の定義及び持続を考慮するプロセスに基づき，使命，ビジョン及び価値観の明確化に参画している．		
		文化を変化させる必要性を考慮するプロセスとともに，現在の文化の理解がある．		
		組織のアイデンティティの変更を，認識した利害関係者に非公式に伝達している．		
	4	組織の文化は，使命，ビジョン及び価値観と一貫している．		

Table A.4 *(continued)*

Subclause	Maturity level		Conclusion	
	Level	Item[a]	YES	Results/ comment[b]
		A clearly defined understanding of the current culture, along with a process for considering the need for a change to that culture, is implemented and maintained.		
		The strategic direction of the organization and its policy are aligned with its mission, vision, values and culture.		
		Changes to any of these identity elements are communicated within the organization and to its interested parties, as appropriate.		
	5	A process for reviewing these elements at planned intervals by top management is well established and maintained. This includes consideration of external and internal issues as part of the verification of alignment between the elements of the identity of the organization, its context, its strategic direction and its policy.		

[a] Items outlined in levels 3 to 5 are intended to be a progression of thought that is based on the guidance provided in the applicable subclause.
[b] This may include recognition of aspects where the organization is partially meeting a maturity level.

表 A.4（続き）

細分箇条	成熟度レベル		結論	
	レベル	項目 a)	YES	結果／コメント b)
		その文化を変化させる必要性を考慮するプロセスとともに，明確に定めた現在の文化の理解を実施し，維持している．		
		組織の戦略的方向性及び組織の方針は，使命，ビジョン，価値観及び文化と一貫している．		
		必要に応じて，組織内で，また，利害関係者にこれらのアイデンティティの要素の変更を伝達している．		
	5	計画した間隔でトップマネジメントがこれらの要素をレビューするプロセスを十分に確立し，維持している．これには，組織のアイデンティティの要素，組織の状況，戦略的方向性及び方針の一貫性を検証する一環として，外部及び内部の課題を考慮することを含む．		

注 a) レベル3からレベル5までに記載している項目は，当該細分箇条に示した手引に基づいており，成熟度のレベルを示すことを意図している．
b) 結果／コメント欄は，組織が一つの成熟度レベルを部分的に満たしているという状況の認識を含んでもよい．

Table A.5 — Self-assessment of the detailed elements of 7.1

Subclause	Maturity level		Conclusion	
	Level	Item[a]	YES	Results/ comment[b]
7.1 Leadership — General	1	Processes for defining, maintaining and communicating the leadership's vision, mission and values, and for promoting an internal environment in which people are engaged and committed to the achievement of the organization's objectives are carried out in an informal or ad hoc manner.		
	2	Key processes, such as those related to establishing the organization's identity, a culture of trust, integrity and teamwork, the necessary resources, training and authority to act, ensuring behavioural attributes are defined, and supporting leadership development, are determined.		
		Only some interrelationships between leadership and commitment, including the maintenance of a competitive organizational structure, maintaining unity of purpose and direction, and reinforcement of values and expectations are determined.		
	3	Processes and the interactions of activities related to the organization's identity, cultural aspects, the provision of resources, training, the authority to act and behavioural factors are accounted for.		

附属書A（参考）

表A.5 － 7.1 に対する自己評価

細分箇条	成熟度レベル		結論	
	レベル	項目 a)	YES	結果／コメント b)
7.1 リーダーシップ一般	1	リーダーシップのビジョン，使命及び価値観を定め，維持し，伝達するプロセス，並びに人々が組織の目標の達成に積極的に参画し，コミットメントする内部環境を促進するためのプロセスを，非公式又はその場限りで実施している．		
	2	組織のアイデンティティの確立，信頼及び誠実の文化，チームワーク，必要な資源，訓練及び権限，行動属性を定めることを確実にすること，並びにリーダーシップ育成支援に関係する，重要なプロセスを決定している．		
		競争力のある組織構造の維持，目的及び方向性の統一の維持，並びに価値観及び期待の補強を含め，リーダーシップとコミットメントとの間では，一部の相互関係しか明確にしていない．		
	3	プロセス並びに組織のアイデンティティ，文化的側面，資源の提供，訓練，権限及び行動要因に関係する活動の相互作用を考慮している．		

Table A.5 *(continued)*

Subclause	Maturity level		Conclusion	
	Level	Item[a]	YES	Results/comment[b]
		A competitive organizational structure and unity of purpose are established.		
		Values and expectations are established and communicated.		
		Leadership development is defined.		
		Processes to maintain the culture and promote accountability are acted on.		
		Maintenance of the organizational structure and unity of purpose in relation to the context of the organization, personally and/or regularly reinforcing values and expectations, are included in process determination.		
	4	Processes and their interactions are systematically determined in such a way that outputs and outcomes are concise and create an internal environment in which people are engaged and committed to the achievement of the organization's objectives, and in a way that promotes understanding and supports the organization's ability to achieve sustained success.		
		All relevant factors and their interrelationships are considered in process determination.		

附属書 A（参考）

表 A.5（続き）

細分箇条	成熟度レベル		結論	
	レベル	項目 a)	YES	結果／コメント b)
		競争力のある組織構造及び目的の統一を確立している．		
		価値観及び期待を確立し，伝達している．		
		リーダーシップの育成を定めている．		
		文化を維持し，説明責任を促進するプロセスを実施している．		
		個人的に及び／又は定期的に価値観及び期待を補強し，組織の状況と関係する，組織構造の維持及び目的の統一を，プロセスの決定に含めている．		
	4	アウトプット及び成果が簡潔であり，持続的成功を達成する組織の能力への理解を促進し，それを支援するような方法で，組織の目標の達成に積極的に参画し，コミットメントする内部環境を生み出すような様式で，プロセス及びその相互作用を体系的に決定している．		
		全ての関連する要因及びその相互関係を，プロセスの決定で考慮している．		

Table A.5 *(continued)*

Subclause	Maturity level		Conclusion	
	Level	Item[a]	YES	Results/comment[b]
	5	Processes and the interactions of leadership with all levels of the organization are dynamically determined and used to establish and sustain the success of the organization.		

[a] Items outlined in levels 3 to 5 are intended to be a progression of thought that is based on the guidance provided in the applicable subclause.
[b] This may include recognition of aspects where the organization is partially meeting a maturity level.

附属書A (参考)

表 A.5 (続き)

細分箇条	成熟度レベル		結論	
	レベル	項目 a)	YES	結果／コメント b)
	5	プロセス及び組織の全ての階層のリーダーシップの相互作用を精力的に決定し，組織の成功を確立し，持続するために活用している．		

注 a) レベル3からレベル5までに記載している項目は，当該細分箇条に示した手引に基づいており，成熟度のレベルを示すことを意図している．

b) 結果／コメント欄は，組織が一つの成熟度レベルを部分的に満たしているという状況の認識を含んでもよい．

Table A.6 — Self-assessment of the detailed elements of 7.2

Subclause	Maturity level		Conclusion	
	Level	Item[a]	YES	Results/ comment[b]
7.2 Policy and strategy	1	Processes for determining the organization's policy and strategy are done in an informal or ad hoc manner.		
	2	The policy and strategy, and the basic strategy framework, are determined.		
	3	Processes and the interactions related to the policy and strategy are defined to address all applicable aspects, models and factors.		
		The organization's identity, the context of the organization and long-term perspective, a competitive profile and consideration of competitive factors are determined.		
		The policy and strategy decisions are reviewed for continuing suitability and changed as deemed necessary by top management.		
	4	Processes and their interactions are systematically determined to ensure the policy and strategy provide a comprehensive framework for process management, to support deployment and facilitate changes, as well as to effectively account for applicable aspects and factors.		

附属書 A (参考)

表 A.6 − 7.2 に対する自己評価

細分箇条	成熟度レベル		結論	
	レベル	項目 a)	YES	結果／コメント b)
7.2 方針及び戦略	1	組織の方針及び戦略を決定するプロセスを，非公式又はその場限りで実行している．		
	2	方針，戦略及び基本戦略枠組みを決定している．		
	3	全ての適用可能な側面，モデル及び要因に取り組むため，方針及び戦略に関係するプロセス及び相互作用を定めている．		
		組織のアイデンティティ，組織の状況及び長期的な展望，組織の能力プロファイル並びに競争的要因の考慮を明確にしている．		
		方針及び戦略に関わる決定を，継続的な適切性のためにレビューし，トップマネジメントが必要と考えた場合には変更している．		
	4	方針及び戦略によって，プロセスのマネジメントの総合的な枠組みを提供することを確実にし，適用可能な側面及び要因を効果的に考慮するだけでなく，展開を支援し，かつ，変化を容易にするため，プロセス及びその相互作用を体系的に決定している．		

Table A.6 *(continued)*

Subclause	Maturity level		Conclusion	
	Level	Item[a]	YES	Results/comment[b]
		The processes for maintaining a standardized or custom model for a strategy framework and the policy are determined and address and aid in the mitigation of risks, while taking advantage of opportunities.		
	5	Processes and the relationships between the policy and strategic direction are dynamically determined, with all applicable aspects and factors accounted for, such that a comprehensive framework exists to support the establishment, maintenance and managing of processes.		
		The needs of all interested parties are addressed and the policy and strategy are utilized to manage the business in a comprehensive way.		
[a] Items outlined in levels 3 to 5 are intended to be a progression of thought that is based on the guidance provided in the applicable subclause. [b] This may include recognition of aspects where the organization is partially meeting a maturity level.				

附属書 A（参考）

表 A.6（続き）

細分箇条	成熟度レベル		結論	
	レベル	項目 a)	YES	結果／コメント b)
		戦略的枠組み及び方針に関する標準化された又はカスタマイズされたモデルを維持するプロセスを決定し，機会を利用しながら，リスクの軽減に取り組み，役立てている．		
	5	全ての適用可能な側面及び要因を考慮しながら，プロセス及び方針と戦略的方向性との関係を精力的に決定しており，それによって，プロセスの確立，維持及びマネジメントを支援する総合的枠組みが存在する．		
		全ての利害関係者のニーズに取り組んでおり，方針及び戦略を利用しながら，事業を総合的な方法でマネジメントしている．		

注 a) レベル3からレベル5までに記載している項目は，当該細分箇条に示した手引に基づいており，成熟度のレベルを示すことを意図している．
 b) 結果／コメント欄は，組織が一つの成熟度レベルを部分的に満たしているという状況の認識を含んでもよい．

Table A.7 — Self-assessment of the detailed elements of 7.3

Subclause	Maturity level		Conclusion	
	Level	Item[a]	YES	Results/comment[b]
7.3 Objectives	1	Processes for determining the organization's objectives are done in an informal or ad hoc manner.		
		Only the short-term objectives are defined.		
	2	Processes for determining the objectives are defined and the objectives show some interrelationship with policy and strategy.		
		The objectives are quantifiable, where possible, but are not clearly understood.		
		Processes and the interactions of short- and long-term objectives with the policy and strategy are defined, including the ability to demonstrate leadership and commitment outside the organization.		
	3	Processes for defining, maintaining and deploying the objectives, including the relationship with the policy and strategy, are in place and maintained, including the need to establish clearly understood and quantifiable short- and long-term objectives that also demonstrate leadership and commitment outside the organization.		

表 A.7 − 7.3 に対する自己評価

細分箇条	成熟度レベル		結論	
	レベル	項目 a)	YES	結果／コメント b)
7.3 目標	1	組織の目標を決定するプロセスを,非公式又はその場限りで実行している.		
		短期的目標しか定めていない.		
	2	目標を決定するプロセスを定めており,その目標は方針及び戦略との何らかの相互関係を示している.		
		目標は,可能な場合には定量化可能であるが,明確に理解されていない.		
		組織外部でリーダーシップ及びコミットメントを実証する能力を含め,プロセス並びに短期的及び長期的目標の方針及び戦略との相互作用を定めている.		
	3	方針及び戦略との関係を含めて,目標を定め,維持し,展開するプロセスを整備し,維持している.これには,組織外部に対してもリーダーシップ及びコミットメントを実証するような,明確に理解され定量化可能な短期的及び長期的目標を確立する必要性を含む.		

Table A.7 *(continued)*

Subclause	Maturity level			Conclusion	
	Level	Item[a]		YES	Results/comment[b]
		The short- and long-term objectives are defined, and the relationship with the policy and strategy is evident.			
	4	Processes and the relationship between the policy, strategy and the demonstrated leadership and commitment outside the organization are dynamically determined and maintained.			
	5	Short- and long-term objectives are quantifiable, clearly understood, deployed and updated to maintain the relationship with the policy and strategy, such that top management's leadership and commitment are demonstrated both internally and outside the organization.			

[a] Items outlined in levels 3 to 5 are intended to be a progression of thought that is based on the guidance provided in the applicable subclause.
[b] This may include recognition of aspects where the organization is partially meeting a maturity level.

附属書 A（参考）

表 A.7（続き）

細分箇条	成熟度レベル			結論	
	レベル	項目 a)		YES	結果／コメント b)
		短期的及び長期的目標を定めており，方針及び戦略との関係が明白である．			
	4	プロセス並びに方針，戦略及び組織外部で実証されたリーダーシップ・コミットメントの間の関係を精力的に決定し，維持している．			
	5	方針及び戦略との関係を維持するため，短期的及び長期的目標を定量化し，明確に理解し，展開し，更新しており，それによって，トップマネジメントのリーダーシップ及びコミットメントを組織内外で実証している．			

注 a) レベル3からレベル5までに記載している項目は，当該細分箇条に示した手引に基づいており，成熟度のレベルを示すことを意図している．
 b) 結果／コメント欄は，組織が一つの成熟度レベルを部分的に満たしているという状況の認識を含んでもよい．

Table A.8 — Self-assessment of the detailed elements of 7.4

Subclause	Maturity level		Conclusion	
	Level	Item[a]	YES	Results/comment[b]
7.4 Communication	1	The processes for communicating the policy, strategy and objectives are done in an informal or ad hoc manner.		
	2	The processes for determining the types and degree of communication needed are defined.		
	3	Communication processes are defined and facilitate meaningful, timely and continual communication tailored to the differing needs of recipients, as it relates to the policy, strategy and relevant objectives.		
		The interrelationships of this communication are clear with regard to the differing needs of recipients and how the policy, strategy and relevant objectives are used to aid in the sustained success of the organization.		
		A feedback mechanism is in place and incorporates provisions to proactively address changes in the organization's context.		

附属書A（参考）

表A.8 − 7.4 に対する自己評価

細分箇条	成熟度レベル		結論	
	レベル	項目 a)	YES	結果／コメント b)
7.4 コミュニケーション	1	方針，戦略及び目標を伝達するプロセスを，非公式又はその場限りで実行している．		
	2	必要となるコミュニケーションの種類及び程度を決定するプロセスを定めている．		
	3	コミュニケーションプロセスを定めており，方針，戦略及び関連する目標に関係している場合には，受け取り側の異なるニーズに合わせて，有意義で，時宜を得た，継続的なコミュニケーションを容易にしている．		
		このコミュニケーションの相互関係は，受け取り側の異なるニーズ，及び組織の持続的成功を支援するため，方針，戦略及び関連する目標を活用する方法に関して明確である．		
		フィードバックの仕組みを整備しており，組織の状況の変化に積極的に取り組むための備えを取り入れている．		

Table A.8 *(continued)*

Subclause	Maturity level		Conclusion	
	Level	Item[a]	YES	Results/comment[b]
	4	The communication processes systematically facilitate communication regarding the policy, strategy and objectives to all relevant interested parties, in support of the organization's sustained success, while also accounting for the need to deploy communication when change is realized.		
		Communication methods show a direct relationship to the context of the organization and the feedback mechanism is well defined and effectively deployed.		
	5	The processes for communicating the policy, strategy and objectives are dynamic, with the interrelationships of the policy, strategy and objectives being clearly conveyed to all recipients, such that the differing needs of each are accounted for.		

[a] Items outlined in levels 3 to 5 are intended to be a progression of thought that is based on the guidance provided in the applicable subclause.
[b] This may include recognition of aspects where the organization is partially meeting a maturity level.

附属書 A（参考）

表 A.8（続き）

細分箇条	成熟度レベル		結論	
	レベル	項目 a)	YES	結果／コメント b)
	4	コミュニケーションプロセスによって，変化が実現した場合にコミュニケーションを展開する必要性をも考慮しながら，組織の持続的成功を支援する，全ての密接に関連する利害関係者への方針，戦略及び目標に関するコミュニケーションを体系的に促進している．		
		コミュニケーション方法は，組織の状況との直接的な関係を示しており，フィードバックの仕組みを十分定め，効果的に展開している．		
	5	方針，戦略及び目標を伝達するプロセスは動的なものであり，方針，戦略及び目標の相互関係が全ての受け取り側に明確に伝達され，それぞれの異なるニーズが考慮されるようになっている．		

注 a) レベル3からレベル5までに記載している項目は，当該細分箇条に示した手引に基づいており，成熟度のレベルを示すことを意図している．

b) 結果／コメント欄は，組織が一つの成熟度レベルを部分的に満たしているという状況の認識を含んでもよい．

Table A.9 — Self-assessment of the detailed elements of 8.1

Subclause	Maturity level		Conclusion	
	Level	Item[a]	YES	Results/comment[b]
8.1 Process management — General	1	Processes are managed in an informal or ad hoc manner.		
	2	Key processes, such as those relating to customer satisfaction and operations related to product and service, are managed.		
		The effectiveness of the processes is individually measured, and acted upon. Interactions between processes are not well managed.		
	3	Processes and their interactions are managed as a system. Interaction conflicts between processes are identified and resolved in a systematic way.		
		Processes are delivering predictable results.		
		Process performance has reached that of average organizations in the sector where the organization operates.		
	4	Process management is integrated with the deployment of the organization's policies, strategy and objectives.		
		The effectiveness and efficiency of processes and their interactions are systematically reviewed and improved.		

附属書 A（参考）

表 A.9 − 8.1 に対する自己評価

細分箇条	成熟度レベル			結論	
	レベル	項目 [a]		YES	結果／コメント [b]
8.1 プロセスのマネジメント――一般	1	プロセスを，非公式又はその場限りでマネジメントしている．			
	2	製品及びサービスに関係する顧客満足及び運用に関するような，重要なプロセスをマネジメントしている．			
		プロセスの有効性を個別に測定し，それに基づいて行動している．プロセス間の相互作用を十分にマネジメントしていない．			
	3	プロセス及びその相互作用をシステムとしてマネジメントしている．プロセス間の相互作用の対立を特定し，体系的な方法で解決している．			
		プロセスによって，予測可能な結果を得ている．			
		プロセスのパフォーマンスは，組織が事業運営する分野での平均的な組織のパフォーマンスに到達している．			
	4	プロセスのマネジメントを，組織の方針，戦略及び目標の展開と統合している．			
		プロセス及びその相互作用の有効性及び効率を，体系的にレビューし，改善している．			

Table A.9 *(continued)*

Subclause	Maturity level		Conclusion	
	Level	Item[a]	YES	Results/comment[b]
		Process performance has exceeded that of average organizations in the sector where the organization operates.		
	5	All relevant processes and their interactions are proactively managed, including outsourced processes, to ensure that they are effective and efficient, in order to achieve the organization's policies, strategy and objectives.		
		Processes and their interactions are adapted and optimized to the context of the organization.		
		Process performance has reached that of leading organizations in the sector where the organization operates.		

[a] Items outlined in levels 3 to 5 are intended to be a progression of thought that is based on the guidance provided in the applicable subclause.
[b] This may include recognition of aspects where the organization is partially meeting a maturity level.

附属書A（参考）

表A.9（続き）

細分箇条	成熟度レベル		結論	
	レベル	項目 [a]	YES	結果／コメント [b]
		プロセスのパフォーマンスは，組織が事業運営する分野での平均的な組織のパフォーマンスを超過している．		
	5	組織の方針，戦略及び目標を達成するために，プロセスが効果的及び効率的であることを確実にするよう，外部委託するプロセスを含め，全ての関連するプロセス及びその相互作用を積極的にマネジメントしている．		
		プロセス及びその相互作用を，組織の状況に合わせて適応し，最適化している．		
		プロセスのパフォーマンスは，組織が事業運営する分野での先導的な組織のパフォーマンスに到達している．		

注 [a] レベル3からレベル5までに記載している項目は，当該細分箇条に示した手引に基づいており，成熟度のレベルを示すことを意図している．
　　[b] 結果／コメント欄は，組織が一つの成熟度レベルを部分的に満たしているという状況の認識を含んでもよい．

Table A.10 — Self-assessment of the detailed elements of 8.2

Subclause	Maturity level		Conclusion	
	Level	Item[a]	YES	Results/comment[b]
8.2 Determination of processes	1	Processes are determined in an informal or ad hoc manner.		
	2	Key processes, such as those relating to customer satisfaction and operations related to product and service, are determined.		
		Interactions between processes are not well determined.		
	3	Processes and their interactions are determined to address not only operations related to product and service, but also provision of resources and managerial activities (e.g. planning, measuring, analysis, improvement).		
		The needs and expectations of identified interested parties are used as inputs into process determination.		
	4	Processes and their interactions are systematically determined to ensure that their outputs continue to meet the needs and expectations of customers and other interested parties.		
		All interested parties are considered in process determination.		

表 A.10 – 8.2 に対する自己評価

細分箇条	成熟度レベル		結論	
	レベル	項目 a)	YES	結果／コメント b)
8.2 プロセスの決定	1	プロセスを，非公式又はその場限りで決定している．		
	2	製品及びサービスに関係する顧客満足及び運用に関するような，重要なプロセスを決定している．		
		プロセス間の相互作用を十分に決定していない．		
	3	製品及びサービスに関係する運用だけでなく，資源の提供及びマネジメント活動（例えば，計画，測定，分析，改善）に取り組むためのプロセス及びその相互作用を決定している．		
		特定した利害関係者のニーズ及び期待を，プロセス決定のインプットとして利用している．		
	4	プロセス及びその相互作用を，アウトプットが顧客及びその他の利害関係者のニーズ及び期待を継続的に満たすことを確実にするよう，体系的に決定している．		
		プロセスの決定において，全ての利害関係者を考慮している．		

Table A.10 *(continued)*

Subclause	Maturity level		Conclusion	
	Level	Item[a]	YES	Results/comment[b]
	5	Processes and their interactions are determined and changed flexibly according to the organization's policies, strategy and objectives.		

[a] Items outlined in levels 3 to 5 are intended to be a progression of thought that is based on the guidance provided in the applicable subclause.
[b] This may include recognition of aspects where the organization is partially meeting a maturity level.

表 A.10 （続き）

細分箇条	成熟度レベル			結論	
	レベル	項目 [a]	YES	結果／コメント [b]
	5	プロセス及びその相互作用を，組織の方針，戦略及び目標に従って決定し，柔軟に変更している．		

注 [a] レベル3からレベル5までに記載している項目は，当該細分箇条に示した手引に基づいており，成熟度のレベルを示すことを意図している．
  [b] 結果／コメント欄は，組織が一つの成熟度レベルを部分的に満たしているという状況の認識を含んでもよい．

Table A.11 — Self-assessment of the detailed elements of 8.3

Subclause	Maturity level		Conclusion	
	Level	Item[a]	YES	Results/comment[b]
8.3 Process responsibility and authority	1	Process responsibilities are defined in an informal or ad hoc manner.		
	2	A process owner is appointed for each process.		
		The competences required for the people associated with the individual processes are not defined.		
	3	For each process, a process owner is appointed, who has defined responsibilities and authorities to establish, maintain, control and improve the process.		
		A policy to avoid and resolve potential disputes in managing processes exists.		
		The competences required for process owners are defined.		
	4	A process owner is appointed for each process, with sufficient responsibility, authority and competence to establish, maintain, control and improve the process and its interaction with other processes.		
		The competences required for the people associated with individual processes are well defined and continually improved.		

表 A.11 － 8.3 に対する自己評価

細分箇条	成熟度レベル			結論	
	レベル	項目 a)		YES	結果／コメント b)
8.3 プロセスの責任及び権限	1	プロセスの責任を，非公式又はその場限りで定めている．			
	2	各プロセスに対して，プロセスオーナを任命している．			
		個々のプロセスに関連する人々に必要な力量を定めていない．			
	3	各プロセスに対して，プロセスを確立し，維持し，管理し，改善するための定義した責任及び権限をもつプロセスオーナを任命している．			
		プロセスのマネジメントにおける潜在的な紛争を回避し，解決するための方針が存在している．			
		プロセスオーナに必要な力量を定めている．			
	4	各プロセスに対して，プロセス及び他のプロセスとの相互作用を確立し，維持し，管理し，改善するために十分な責任，権限及び力量を伴ったプロセスオーナを任命している．			
		個々のプロセスに関連する人々に必要な力量を十分に定めており，継続的に改善している．			

Table A.11 *(continued)*

Subclause	Maturity level		Conclusion	
	Level	Item[a]	YES	Results/ comment[b]
	5	Responsibilities, authorities and roles of process owners are recognized throughout the organization.		
		Responsibilities and authorities for interactions between processes are well defined.		
		The people associated with individual processes have sufficient competences for the tasks and activities involved.		

[a] Items outlined in levels 3 to 5 are intended to be a progression of thought that is based on the guidance provided in the applicable subclause.
[b] This may include recognition of aspects where the organization is partially meeting a maturity level.

附属書 A (参考)

表 A.11 (続き)

細分箇条	成熟度レベル			結論	
	レベル	項目 a)	YES	結果／コメント b)
	5	プロセスオーナの責任，権限及び役割を，組織全体を通じて認識している．		
		プロセス間の相互作用に関する責任及び権限を十分に定めている．		
		個々のプロセスに関連する人々は，関与する業務及び活動のための十分な力量をもっている．		

注 a) レベル3からレベル5までに記載している項目は，当該細分箇条に示した手引に基づいており，成熟度のレベルを示すことを意図している．
b) 結果／コメント欄は，組織が一つの成熟度レベルを部分的に満たしているという状況の認識を含んでもよい．

Table A.12 — Self-assessment of the detailed elements of 8.4 (alignment/linkage)

Subclause	Maturity level		Conclusion	
	Level	Item[a]	YES	Results/comment[b]
8.4 Managing processes (managing alignment/linkage between the processes)	1	Processes are aligned and linked in an informal or ad hoc manner.		
	2	Alignment/linkage between processes is discussed but major concerns of managers are on individual processes.		
	3	The network of processes, their sequence and interactions are visualized in a graphic to understand the roles of each process in the system and its effects on the performance of the system.		
		Processes and their interactions are managed as a system to enhance alignment/linkage between the processes.		
	4	Criteria for the outputs of processes are determined. The capability and performance of processes are evaluated and improved.		
		The risks and opportunities associated with processes are assessed and necessary actions are implemented to prevent, detect or mitigate undesired events.		
		Processes and their interactions are reviewed on a regular basis and suitable actions are taken for their improvement to support sustained and effective processes.		

附属書 A（参考）

表 A.12 − 8.4（すり合わせ・連携）に対する自己評価

細分箇条	成熟度レベル		結論	
	レベル	項目 a)	YES	結果／コメント b)
8.4 プロセスのマネジメント（プロセス間のすり合わせ・連携のマネジメント）	1	プロセスを，非公式又はその場限りですり合わせ，連携している．		
	2	プロセス間のすり合わせ・連携を議論しているが，管理者の主な懸念は個々のプロセスに関するものである．		
	3	システム内における各プロセスの役割及びそのシステムのパフォーマンスへの影響を理解するため，プロセスのネットワーク，その順序及び相互作用を図によって視覚化している．		
		プロセス間のすり合わせ・連携を強化するために，プロセス及びその相互作用を一つのシステムとしてマネジメントしている．		
	4	プロセスのアウトプットに対する基準を決定している．プロセスの実現能力及びパフォーマンスを評価し，改善している．		
		プロセスに関連するリスク及び機会を評価し，望ましくない事象を防止し，検出し，軽減するために必要な処置を実施している．		
		持続的かつ効果的なプロセスを支援するために，プロセス及びその相互作用を定期的にレビューし，それらの改善のための適切な処置を実施している．		

Table A.12 *(continued)*

Subclause	Maturity level		Conclusion	
	Level	Item[a]	YES	Results/comment[b]
	5	The capability and performance of processes are sufficient to effectively and efficiently achieve the performance expected by the system.		
		Cross-functional teams or committees under top management's leadership facilitate their review and improvement of the processes.		

[a] Items outlined in levels 3 to 5 are intended to be a progression of thought that is based on the guidance provided in the applicable subclause.
[b] This may include recognition of aspects where the organization is partially meeting a maturity level.

表 A.12（続き）

細分箇条	成熟度レベル		結論	
	レベル	項目 a)	YES	結果／コメント b)
	5	プロセスの実現能力及びパフォーマンスは，システムによって期待されるパフォーマンスを効果的かつ効率的に達成するために十分である．		
		トップマネジメントのリーダーシップによる，機能横断チーム又は委員会が，プロセスのレビュー及び改善を容易にしている．		

注 a) レベル3からレベル5までに記載している項目は，当該細分箇条に示した手引に基づいており，成熟度のレベルを示すことを意図している．
 b) 結果／コメント欄は，組織が一つの成熟度レベルを部分的に満たしているという状況の認識を含んでもよい．

Table A.13 — Self-assessment of the detailed elements of 8.4 (attaining a higher level of performance)

Subclause	Maturity level		Conclusion	
	Level	Item[a]	YES	Results/ comment[b]
8.4 Managing processes (attaining a higher level of performance)	1	Processes and their interactions are improved in an informal or ad hoc manner.		
	2	Improvement of processes and their interactions are loosely related with the organization's policies, strategy and objectives.		
	3	Processes and their interactions are improved based on the organization's policies, strategy and objectives.		
		The achievement of the objectives for improvement of processes and their interactions are reviewed on a regular basis.		
	4	Processes and their interactions are systematically improved to achieve the organization's policies, strategy and objectives.		
		The action plans for attaining the objectives are determined, taking into account the resources needed and their availability.		
		People are motivated to engage in the improvement activities and propose opportunities for improvement in the processes for which they are in charge.		

附属書A（参考）

表A.13 － 8.4（より高いパフォーマンスの達成）に対する自己評価

細分箇条	成熟度レベル		結論	
	レベル	項目 a)	YES	結果／コメント b)
8.4 プロセスのマネジメント (より高いパフォーマンスの達成)	1	プロセス及びその相互作用を，非公式又はその場限りで改善している．		
	2	プロセス及びその相互作用の改善は，組織の方針，戦略及び目標と緩やかに関連している．		
	3	プロセス及びその相互作用を，組織の方針，戦略及び目標に基づいて改善している．		
		プロセス及びその相互作用に関する改善目標の達成を，定期的にレビューしている．		
	4	組織の方針，戦略及び目標を達成するため，プロセス及びその相互作用を体系的に改善している．		
		目標を達成するための実施計画を，必要な資源及びその利用可能性を考慮しながら決定している．		
		人々が改善活動に積極的に参加し，自分が関わっているプロセスにおいて改善の機会を提案するよう動機付けている．		

Table A.13 *(continued)*

Subclause	Maturity level		Conclusion	
	Level	Item[a]	YES	Results/ comment[b]
	5	Improvement of processes and their interaction are dynamically managed through the organization's policies, strategy and objectives.		
		The need to develop or acquire new technologies, or develop new products and services or their features, for added value, are considered.		
		The achievement of the objectives for improvement, the progress of the action plans, and the effects on the related organization's policies, strategy and objectives are reviewed on a regular basis, and necessary corrective actions are taken.		

[a] Items outlined in levels 3 to 5 are intended to be a progression of thought that is based on the guidance provided in the applicable subclause.
[b] This may include recognition of aspects where the organization is partially meeting a maturity level.

附属書A（参考）

表A.13（続き）

細分箇条	成熟度レベル		結論	
	レベル	項目 a)	YES	結果／コメント b)
	5	プロセス及びその相互作用の改善を，組織の方針，戦略及び目標を通じて動的にマネジメントしている．		
		新しい技術を開発若しくは獲得する必要性，又は付加価値を付ける新しい製品及びサービス若しくはその特徴を開発する必要性を考慮している．		
		改善目標の達成，実施計画の進捗状況並びに関係する組織の方針，戦略及び目標への影響について定期的にレビューし，必要な是正処置をとっている．		

注 a) レベル3からレベル5までに記載している項目は，当該細分箇条に示した手引に基づいており，成熟度のレベルを示すことを意図している．
    b) 結果／コメント欄は，組織が一つの成熟度レベルを部分的に満たしているという状況の認識を含んでもよい．

Table A.14 — Self-assessment of the detailed elements of 8.4 (maintaining the level)

Subclause	Maturity level		Conclusion	
	Level	Item[a]	YES	Results/comment[b]
8.4 Managing processes (maintaining the level attained)	1	Processes and their interactions are operated in an informal or ad hoc manner.		
	2	Procedures are determined for relevant processes but not well followed.		
		Deviations are no concern of process owners.		
	3	Procedures are determined for each process, including the criteria for its outputs and operational conditions.		
		Consideration is given to education and training.		
		Managers take necessary corrective actions when the procedures are not followed.		
		Resources necessary for people to follow the procedures are made available.		
		Processes are monitored on a regular basis to detect deviations.		
	4	Procedures ensure conformance of the outputs to the criteria.		
		People have sufficient knowledge and skills to follow the procedures and understand the impacts of not following the procedures.		
		Consideration is given to motivation and human error prevention.		

附属書 A（参考）

表 A.14 − 8.4（レベルの維持）に対する自己評価

細分箇条	成熟度レベル		結論	
	レベル	項目 a)	YES	結果／コメント b)
8.4 プロセスのマネジメント（達成されたレベルの維持）	1	プロセス及びその相互作用を，非公式又はその場限りで運用している．		
	2	関連するプロセスに対して手順を明確にしているが，十分に従っていない．		
		逸脱は，プロセスオーナにとっての関心事ではない．		
	3	プロセスのアウトプット及び運用条件に対する基準を含め，各プロセスに対して手順を明確にしている．		
		教育及び訓練への配慮が行われている．		
		手順に従っていない場合には，管理者が必要な是正処置をとっている．		
		人々が手順に従うために必要な資源を提供している．		
		逸脱を検出するため，プロセスを定期的に監視している．		
	4	手順によって，基準へのアウトプットの適合を確実にしている．		
		人々は手順に従うために十分な知識及び技能を備えており，手順に従わないことによる影響について理解している．		
		動機付け及び人的ミスの防止への配慮を行っている．		

Table A.14 *(continued)*

Subclause	Maturity level		Conclusion	
	Level	Item[a]	YES	Results/comment[b]
		Check points and related performance indicators are determined to detect deviations (which are mainly caused by changes in people, equipment, methods, material, measurement and environment for operation of processes) and to take appropriate actions when necessary.		
	5	A system for determining the knowledge and skills needed for each process, evaluating the knowledge and skills of the process operators, and providing qualifications for operating the process is established.		
		People are engaged in the development or revision of the procedures.		
		Risks and opportunities in the procedures are identified, assessed and reduced by improving the procedures.		
		Changes in processes are clarified and shared to prevent deviations.		

[a] Items outlined in levels 3 to 5 are intended to be a progression of thought that is based on the guidance provided in the applicable subclause.
[b] This may include recognition of aspects where the organization is partially meeting a maturity level.

附属書A（参考）

表 A.14（続き）

細分箇条	成熟度レベル		結論	
	レベル	項目 a)	YES	結果／コメント b)
		（主に人々，設備，方法，材料，測定及びプロセス運用環境における変化によって引き起こされる）逸脱を検出し，必要な場合には適切な処置を実施するためのチェックポイント及び関係するパフォーマンス指標を明確にしている．		
	5	各プロセスに必要な知識及び技能を明確にし，プロセスを運用する人の知識及び技能を評価し，並びにプロセス運用の資格を認定するためのシステムを確立している．		
		人々が手順の開発又は改訂に積極的に参加している．		
		手順におけるリスク及び機会を特定し，評価し，その手順を改善することによって低減している．		
		逸脱を防止するために，プロセスの変更を明確にし，共有している．		

注 a) レベル3からレベル5までに記載している項目は，当該細分箇条に示した手引に基づいており，成熟度のレベルを示すことを意図している．
  b) 結果／コメント欄は，組織が一つの成熟度レベルを部分的に満たしているという状況の認識を含んでもよい．

Table A.15 — Self-assessment of the detailed elements of 9.1

Subclause	Maturity level		Conclusion	
	Level	Item[a]	YES	Results/comment[b]
9.1 Resource management — General	1	Processes to manage the resources to support the operation in an organization are performed in an informal and ad hoc manner.		
		Some of the resource management processes to support the achievement of objectives are determined.		
	2	Support for the effective and efficient use of resources is defined in a limited manner.		
		A basic approach for considering risks and opportunities, including the effects of not having sufficient resources in a timely manner, is in place.		
		Key processes to determine and manage the resources needed for the achievement of its objectives are determined.		
		The efficient and effective uses of resources are not well defined.		
		Resource management processes and the interactions for getting and assigning resources, aligned with the organizational objectives, are present.		
	3	Some processes include an approach for effectively and efficiently applying resources.		

附属書 A（参考）

表 A.15 − 9.1 に対する自己評価

細分箇条	成熟度レベル		結論	
	レベル	項目 a)	YES	結果／コメント b)
9.1 資源のマネジメント―一般	1	組織における運用を支援する資源をマネジメントするプロセスを，非公式及びその場限りで実行している．		
		目標の達成を支援するための，幾つかの資源のマネジメントプロセスが，明確になっている．		
	2	資源の効果的かつ効率的な利用のための支援は，限定的に定められている．		
		十分な資源を適切な時期に保有していない場合の影響も含め，リスク及び機会を考慮するための基本的なアプローチを整備している．		
		組織の目標の達成に必要な資源を明確にし，マネジメントする重要なプロセスが明確である．		
		資源の効果的及び効率的な利用を，十分に定めていない．		
		組織の目標と一貫性のある資源の獲得及び割当てのための，資源のマネジメントプロセス及び相互作用が存在している．		
	3	幾つかのプロセスは，効果的及び効率的に資源を適用するアプローチを含む．		

Table A.15 *(continued)*

Subclause	Maturity level		Conclusion	
	Level	Item[a]	YES	Results/comment[b]
		Resource management processes and the interactions for getting and assigning resources, aligned with the organizational objectives, are systematically implemented.		
	4	Controls to support the effective and efficient use of resources in all processes are established.		
		The accessibility of externally provided resources is confirmed by the organization.		
		External providers are encouraged by the organization to implement improvements on the efficient and effective use of resources.		
		A strategic planning process for getting and assigning resources is in place and is aligned with the organizational objectives in order to achieve effective and efficient performance in support of the sustained success.		
	5	The use of externally provided resources shows continual improvement.		
		There are joint initiatives with external providers to evaluate and incorporate improvements and promote innovations on the use of resources.		

[a] Items outlined in levels 3 to 5 are intended to be a progression of thought that is based on the guidance provided in the applicable subclause.
[b] This may include recognition of aspects where the organization is partially meeting a maturity level.

附属書A（参考）

表 A.15（続き）

細分箇条	成熟度レベル		結論	
	レベル	項目 a)	YES	結果／コメント b)
		組織の目標と一貫性のある資源の獲得及び割当てのための，資源のマネジメントプロセス及び相互作用を体系的に実施している．		
	4	あらゆるプロセスにおいて，効果的及び効率的な資源の利用を支援するための管理を確立している．		
		組織は，外部から提供される資源の入手可能性を確認している．		
		組織は，外部提供者に対し，効果的及び効率的な資源利用の改善策を実施するよう奨励している．		
		資源の獲得及び割当てのための戦略的な計画策定プロセスを整備しており，持続的成功における効果的及び効率的なパフォーマンスの達成のための組織の目標との一貫性がある．		
	5	外部から提供される資源の利用を継続的に改善している．		
		資源の利用を評価し，改善に取り組み，革新を促進する，外部提供者との共同の取組みが存在している．		

注 a) レベル3からレベル5までに記載している項目は，当該細分箇条に示した手引に基づいており，成熟度のレベルを示すことを意図している．
 b) 結果／コメント欄は，組織が一つの成熟度レベルを部分的に満たしているという状況の認識を含んでもよい．

Table A.16 — Self-assessment of the detailed elements of 9.2

Subclause	Maturity level		Conclusion	
	Level	Item[a]	YES	Results/comment[b]
9.2 People	1	Competent, engaged, empowered and motivated people are considered to be a resource in an informal or ad hoc manner.		
		Competence development is provided in an informal or ad hoc manner.		
	2	Processes to attract competent, engaged, empowered and motivated people are implemented.		
		Processes for determining, developing, evaluating and improving resources are evident in some cases.		
		Some competence reviews have been implemented.		
	3	A planned, transparent, ethical and socially responsible approach is applied at all levels throughout the organization.		
		Reviews and evaluations of the effectiveness of actions taken ensure the personal competences (in both the short and long term) are in accordance with the mission, vision and objectives.		
	4	Information, knowledge and experience are shared to provide personal growth.		
		Learning, knowledge transfer and teamwork within the organization are evident.		

表A.16 − 9.2 に対する自己評価

細分箇条	成熟度レベル		結論	
	レベル	項目 a)	YES	結果／コメント b)
9.2 人々	1	力量があり，積極的に参加し，権限委譲され，動機付けられた人々は資源であると，非公式又はその場限りでみなされている．		
		力量の開発を，非公式又はその場限りで提供している．		
	2	力量があり，積極的に参加し，権限委譲され，動機付けられた人々を引き付けるプロセスを実施している．		
		資源を明確にし，開発し，評価し，改善するためのプロセスが，幾つかの場合で明らかである．		
		幾つかの力量のレビューを実施している．		
	3	計画的で，透明で，倫理的で，社会的責任を果たすアプローチを組織全体の全ての階層において適用している．		
		(短期的及び長期的に) 個人の力量を確実にするためにとった処置の有効性についてのレビュー及び評価は，使命，ビジョン及び目標に従っている．		
	4	情報，知識及び経験が，個人の成長をもたらすよう共有されている．		
		学習，知識の移転及び組織内のチームワークは，明らかである．		

Table A.16 *(continued)*

Subclause	Maturity level		Conclusion	
	Level	Item[a]	YES	Results/comment[b]
		Competence development is provided to develop skills for creativity and improvement.		
		People are aware of their personal competences and where they can best contribute to the organization's improvement.		
		Career planning is well developed.		
	5	The results achieved for competent, engaged, empowered and motivated people are shared and compare well with other organizations.		
		People across the organization participate in the development of new processes.		
		Best practices are recognized.		

[a] Items outlined in levels 3 to 5 are intended to be a progression of thought that is based on the guidance provided in the applicable subclause.
[b] This may include recognition of aspects where the organization is partially meeting a maturity level.

表 A.16（続き）

細分箇条	成熟度レベル		結論	
	レベル	項目 a)	YES	結果／コメント b)
		創造性及び改善を図る技能を開発するための力量開発が行われている．		
		人々は，個人の力量，及びどこで自身が組織の改善に最もよく貢献できるかを認識している．		
		キャリアプランを的確に策定している．		
	5	力量があり，積極的に参加し，権限委譲され，動機付けられた人々において達成した結果を共有し，他の組織と十分に比較している．		
		組織全体にわたり，人々が新たなプロセスの開発に参加している．		
		ベストプラクティスが認知されている．		

注 a) レベル3からレベル5までに記載している項目は，当該細分箇条に示した手引に基づいており，成熟度のレベルを示すことを意図している．
 b) 結果／コメント欄は，組織が一つの成熟度レベルを部分的に満たしているという状況の認識を含んでもよい．

Table A.17 — Self-assessment of the detailed elements of 9.3

Subclause	Maturity level		Conclusion	
	Level	Item[a]	YES	Results/comment[b]
9.3 Organizational knowledge	1	Processes to capture the current baseline of organizational knowledge are informal or ad hoc.		
		Processes to develop organizational knowledge are informal or ad hoc.		
	2	Some processes for maintaining and protecting documented organizational knowledge are in place.		
	3	Activities to determine whether explicit or tacit knowledge exists are in place and some are documented.		
		Processes for identifying important information and ensuring the effective distribution of such information throughout the life cycle(s) of relevant products and/or services exist.		
	4	Processes exist for gathering and analysing competitive data.		
		Processes to evaluate staff understanding of relevant organizational knowledge are present.		
		Methods for process owners to evaluate the competencies of utilizing processes exist.		
		Methods for determining and communicating the roles of process owners in managing staff are present.		

表A.17 − 9.3 に対する自己評価

細分箇条	成熟度レベル		結論	
	レベル	項目 a)	YES	結果／コメント b)
9.3 組織の知識	1	組織の知識についての現在のベースラインを取得するプロセスは,非公式又はその場限りである.		
		組織の知識を開発するプロセスは,非公式又はその場限りである.		
	2	文書化した組織の知識を維持し,保護するための幾つかのプロセスを整備している.		
	3	形式知又は暗黙知が存在するかどうかを明確にする活動があり,その幾つかを文書化している.		
		関連する製品及び／又はサービスのライフサイクル全体を通じて,重要な情報を特定し,そうした情報の効果的な配布を確実にするプロセスがある.		
	4	競争データを収集し,分析するプロセスがある.		
		関連する組織の知識のスタッフの理解を評価するプロセスがある.		
		プロセスの利用に関する力量をプロセスオーナが評価する方法が存在する.		
		スタッフのマネジメントにおいてプロセスオーナの役割を明確にし,伝達する方法が存在する.		

Table A.17 *(continued)*

Subclause	Maturity level		Conclusion	
	Level	Item[a]	YES	Results/comment[b]
	5	There are processes for gathering and analysing data from interested parties.		

[a] Items outlined in levels 3 to 5 are intended to be a progression of thought that is based on the guidance provided in the applicable subclause.
[b] This may include recognition of aspects where the organization is partially meeting a maturity level.

附属書A（参考）

表A.17（続き）

細分箇条	成熟度レベル		結論	
	レベル	項目 a)	YES	結果／コメント b)
	5	利害関係者からデータを収集し，分析するプロセスがある．		

注 a) レベル3からレベル5までに記載している項目は，当該細分箇条に示した手引に基づいており，成熟度のレベルを示すことを意図している．
  b) 結果／コメント欄は，組織が一つの成熟度レベルを部分的に満たしているという状況の認識を含んでもよい．

Table A.18 — Self-assessment of the detailed elements of 9.4

Subclause	Maturity level		Conclusion	
	Level	Item[a]	YES	Results/comment[b]
9.4 Technology	1	Advances in current technology are done in an informal or ad hoc manner.		
	2	Some of the processes for identifying the latest innovations and technological developments exist within the organization or the sectors to which it directly or indirectly relates.		
	3	Processes for evaluating the benefits, risks and opportunities for the identified innovations and emergent technologies are in place and support the suitability of product and/or service strategies.		
		Processes for estimating the cost/benefits for adopting suitable innovations and/or emergent technologies are in place.		
		Processes for evaluating the marketability of selected innovations and/or emerging technologies are in place.		
	4	The organizational knowledge and resource capability needed to adapt to the innovations and/or technological changes or advancements are in place.		

附属書 A（参考）

表A.18 − 9.4 に対する自己評価

細分箇条	成熟度レベル		結論	
	レベル	項目 a)	YES	結果／コメント b)
9.4 技術	1	現在の技術における進歩は，非公式又はその場限りである．		
	2	幾つかの最近の革新及び技術開発を特定するプロセスが，組織内又は直接的若しくは間接的に関連する分野に存在している．		
	3	便益，リスク及び機会，又は特定した革新及び新興の技術を評価するプロセスを整備していて，製品及び／又はサービス戦略の適切性を支援している．		
		適切な革新及び／又は新興の技術を採用するためのコスト・便益を見積もるプロセスを整備している．		
		選定した革新及び／又は新興の技術に関わる市場性を評価するプロセスがある．		
	4	革新及び／若しくは技術的変化又は進歩を適応するために必要な，組織の知識及び資源の実現能力がある．		

Table A.18 *(continued)*

Subclause	Maturity level		Conclusion	
	Level	Item[a]	YES	Results/comment[b]
		Processes for evaluating the risks and opportunities for adopting the selected innovations and/or technological changes or advancements are in place.		
	5	Processes for considering the needs of interested parties and offering a suite of innovations as solutions to meet customer expectations are in place.		
		The organization takes measures to keep informed of new technologies and methodologies and their possible benefits.		
		The impact of new technologies and new practices is monitored and evaluated regularly, taking into account internal and external effects, including interested parties and the environment.		

[a] Items outlined in levels 3 to 5 are intended to be a progression of thought that is based on the guidance provided in the applicable subclause.
[b] This may include recognition of aspects where the organization is partially meeting a maturity level.

表 A.18（続き）

細分箇条	成熟度レベル			結論	
	レベル	項目 a)		YES	結果／コメント b)
		選定した革新及び／若しくは技術的変化又は進歩の採用に関わるリスク及び機会を評価するプロセスがある．			
	5	利害関係者のニーズを考慮し，顧客の期待を満たすための解決策として一連の革新を提供するプロセスがある．			
		組織は，新しい技術及び方法論並びにそれらのあり得る便益についての情報を取得し続けるための対策をとっている．			
		利害関係者及び環境を含む，内部及び外部への影響を考慮に入れながら，新しい技術及び新しい実践の影響を定期的に監視し，評価している．			

注 a) レベル3からレベル5までに記載している項目は，当該細分箇条に示した手引に基づいており，成熟度のレベルを示すことを意図している．
b) 結果／コメント欄は，組織が一つの成熟度レベルを部分的に満たしているという状況の認識を含んでもよい．

Table A.19 — Self-assessment of the detailed elements of 9.5

Subclause	Maturity level		Conclusion	
	Level	Item[a]	YES	Results/comment[b]
9.5 Infrastructure and work environment	1	Infrastructure and work environment needs are addressed in an informal or ad hoc manner.		
	2	Some processes for addressing infrastructure and work environment needs are in place.		
	3	Processes that address applicable risks and opportunities and that implement activities for the determination, allocation, provision, measurement or monitoring, improvement, maintenance and protection of the infrastructure and work environment are in place.		
	4	Processes implementing advanced techniques to improve performance and ensure the maximum efficiency in the use of infrastructure and work environment resources are in place.		
		These processes operate in a proactive manner and contribute to the achievement of the organization's objectives, including the fulfilment of statutory and regulatory requirements.		

附属書A (参考)

表A.19 − 9.5 に対する自己評価

細分箇条	成熟度レベル			結論	
	レベル	項目 a)		YES	結果／コメント b)
9.5 インフラストラクチャ及び作業環境	1	インフラストラクチャ及び作業環境に関するニーズに，非公式又はその場限りで取り組んでいる．			
	2	インフラストラクチャ及び作業環境に関するニーズに取り組む幾つかのプロセスを整備している．			
	3	適用可能なリスク及び機会に取り組み，インフラストラクチャ及び作業環境を明確にし，配分し，提供し，測定し，又は監視し，改善し，維持し保護する活動を実施するプロセスを整備している．			
	4	パフォーマンスを改善する高度手法を実施し，インフラストラクチャ及び作業環境資源の利用において最大限の効率を確実にするプロセスを整備している．			
		これらのプロセスは，積極的に機能していて，法令・規制要求事項の遂行を含む，組織の目標の達成に貢献している．			

Table A.19 *(continued)*

Subclause	Maturity level		Conclusion	
	Level	Item[a]	YES	Results/comment[b]
	5	The way in which infrastructure and the work environment are managed becomes a key contributor in the achievement of desired results.		

[a] Items outlined in levels 3 to 5 are intended to be a progression of thought that is based on the guidance provided in the applicable subclause.
[b] This may include recognition of aspects where the organization is partially meeting a maturity level.

表A.19（続き）

細分箇条	成熟度レベル		結論	
	レベル	項目 a)	YES	結果／コメント b)
	5	インフラストラクチャ及び作業環境のマネジメント方法が望ましい結果の達成における主要な寄与因子となる．		

注 a) レベル3からレベル5までに記載している項目は，当該細分箇条に示した手引に基づいており，成熟度のレベルを示すことを意図している．

b) 結果／コメント欄は，組織が一つの成熟度レベルを部分的に満たしているという状況の認識を含んでもよい．

Table A.20 — Self-assessment of the detailed elements of 9.6

Subclause	Maturity level		Conclusion	
	Level	Item[a]	YES	Results/comment[b]
9.6 Externally provided resources	1	The concept of partnership with external providers is informal or ad hoc.		
		External providers are seen as transient and the organization sees no value in developing relationships.		
	2	There is a limited understanding regarding the value in having regular external providers that will deliver a consistent supply.		
	3	There is a good working relationship between the two organizations, with regular communications about issues relating to the product or service provided taking place.		
	4	Top management is committed to developing a close relationship with external providers, with action to develop this at the level of middle management, where close cooperation is carried out.		
		Some project coordination is carried out as it relates to specific product realization processes.		
	5	Both organizations fully appreciate the value of the relationship.		
		There is close interaction between top management staff in the two organizations.		

附属書A（参考）

表A.20 — 9.6 に対する自己評価

細分箇条	成熟度レベル		結論	
	レベル	項目 a)	YES	結果／コメント b)
9.6 外部から提供される資源	1	外部提供者とのパートナシップの概念についての理解は，非公式又はその場限りのものである．		
		外部提供者を一時的なものとみなしていて，組織では関係構築に価値を置いていない．		
	2	整合性のある供給を行う正規の外部提供者を保有することの価値について，理解が限られている．		
	3	双方の組織の間には良好な業務関係があり，提供される製品又はサービスの供給に関連して発生している課題について，定期的なコミュニケーションを行っている．		
	4	トップマネジメントは，外部提供者との緊密な関係の構築をコミットメントしており，これを構築する処置は中級管理者のレベルであり，そこでは密接な協力が行われている．		
		特定の製品の実現プロセスに関連して，一部のプロジェクト調整を実施している．		
	5	双方の組織とも，関係がもつ価値を十分理解している．		
		双方の組織のトップマネジメント要員の間には緊密な相互作用がある．		

Table A.20 *(continued)*

Subclause	Maturity level		Conclusion	
	Level	Item[a]	YES	Results/comment[b]
		There is sharing of some relevant sensitive commercial information.		
		Both organizations engage in business development projects of common interest.		

[a] Items outlined in levels 3 to 5 are intended to be a progression of thought that is based on the guidance provided in the applicable subclause.
[b] This may include recognition of aspects where the organization is partially meeting a maturity level.

表A.20（続き）

細分箇条	成熟度レベル		結論	
	レベル	項目 a)	YES	結果／コメント b)
		何らかの関連する機密性の高い商業情報を共有している．		
		双方の組織とも，共通の利益をもつ事業開発プロジェクトに積極的に参加している．		

注 a) レベル3からレベル5までに記載している項目は，当該細分箇条に示した手引に基づいており，成熟度のレベルを示すことを意図している．
b) 結果／コメント欄は，組織が一つの成熟度レベルを部分的に満たしているという状況の認識を含んでもよい．

Table A.21 — Self-assessment of the detailed elements of 9.7

Subclause	Maturity level		Conclusion	
	Level	Item[a]	YES	Results/comment[b]
9.7 Natural resources	1	There is no managing of natural resources.		
		The organization uses natural resources as required by their processes without considering the potential impacts on their products and services in the future.		
	2	The organization implements some good practices in its current application and use of natural resources.		
	3	The managing of natural resources is aligned within the organization's management system strategy. There is some evidence of improving the actual use of, and minimization of the potential impact of the use of, natural resources.		
	4	The organization recognizes its responsibility to society for managing natural resources.		
		The organization has implemented some best practices in its current application of natural resources.		
	5	The organization recognizes its responsibility to society for managing natural resources related to the life cycle of products and services.		
		The managing of natural resources is widespread in all the areas of the organization.		

表 A.21 — 9.7 に対する自己評価

細分箇条	成熟度レベル		結論	
	レベル	項目 a)	YES	結果／コメント b)
9.7 天然資源	1	天然資源のマネジメントが行われていない．		
		組織は，将来における製品及びサービスへの潜在的な影響を考慮することなく，そのプロセスによって必要となる場合に天然資源を利用している．		
	2	組織は，その現在の天然資源の適用及び利用において幾つかの優れた実践を実施している．		
	3	天然資源のマネジメントが，組織のマネジメントシステム戦略内で一貫している． 天然資源の実際の利用を改善し，天然資源の利用による潜在的な影響を最小限に抑えているという証拠が幾らかある．		
	4	組織は，天然資源のマネジメントにおける組織の社会への責任を認識している．		
		組織は，組織の現在の天然資源の適用において幾つかのベストプラクティスを実施している．		
	5	組織は，製品及びサービスのライフサイクルに関連する天然資源のマネジメントにおけるその社会への責任を認識している．		
		天然資源のマネジメントが，組織の全分野において広範囲に及んでいる．		

Table A.21 *(continued)*

Subclause	Maturity level		Conclusion	
	Level	Item[a]	YES	Results/comment[b]
		The organization addresses both current and future use of natural resources required by its processes.		
		The organization is aware of new trends and technologies for the efficient use of natural resources, and in relation to the needs and expectations of interested parties.		

[a] Items outlined in levels 3 to 5 are intended to be a progression of thought that is based on the guidance provided in the applicable subclause.
[b] This may include recognition of aspects where the organization is partially meeting a maturity level.

附属書 A（参考）

表 A.21（続き）

細分箇条	成熟度レベル			結論	
	レベル	項目 a)		YES	結果／コメント b)
		組織は，組織のプロセスが必要とする天然資源の現在及び今後双方の利用に取り組んでいる．			
		組織は，天然資源の効率的な利用及び利害関係者のニーズ及び期待に関する，新しい傾向及び技術を認識している．			

注 a) レベル3からレベル5までに記載している項目は，当該細分箇条に示した手引に基づいており，成熟度のレベルを示すことを意図している．
  b) 結果／コメント欄は，組織が一つの成熟度レベルを部分的に満たしているという状況の認識を含んでもよい．

Table A.22 — Self-assessment of the detailed elements of 10.1

Subclause	Maturity level		Conclusion	
	Level	Item[a]	YES	Results/comment[b]
10.1 Analysis and evaluation of an organization's performance — General	1	The necessity for updating and understanding the organization's context, policies, strategy and objectives is determined in an informal or ad hoc manner.		
	2	Some information on the organization's performance, the status of its internal activities and resources, changes in its external and internal issues, and the needs and expectations of the interested parties is collected and analysed to update and understand the organization's context, policies, strategy and objectives.		
	3	Available information is collected to update and understand the organization's context, policies, strategy and objectives in a planned manner.		
	4	Based on comprehensive analysis and reviews of available information, the necessity for updating and understanding of the organization's context, policies, strategy and objectives is determined.		

附属書A(参考)

表A.22 − 10.1 に対する自己評価

細分箇条	成熟度レベル		結論	
	レベル	項目 a)	YES	結果／コメント b)
10.1 組織のパフォーマンスの分析及び評価――一般	1	組織の状況,方針,戦略及び目標を更新し,理解する必要性を,非公式又はその場限りの方法で決定している.		
	2	組織の状況,方針,戦略及び目標を更新し,理解するために,組織のパフォーマンス,内部活動及び資源の状態,外部及び内部の課題における変化,並びに利害関係者のニーズ及び期待に関する幾つかの情報を収集し,分析している.		
	3	組織の状況,方針,戦略及び目標を更新し,理解するため,利用可能な情報を計画的に収集している.		
	4	利用可能な情報の総合的な分析及びレビューに基づき,組織の状況,方針,戦略及び目標を更新し,理解する必要性を明確にしている.		

Table A.22 *(continued)*

Subclause	Maturity level		Conclusion	
	Level	Item[a]	YES	Results/comment[b]
	5	A systematic approach is established to collect, analyse and review available information and to determine the necessity for updating and understanding of the organization's context, policies, strategy and objectives, and to identify opportunities for improvement, learning and innovation of the organization's leadership activities.		
[a] Items outlined in levels 3 to 5 are intended to be a progression of thought that is based on the guidance provided in the applicable subclause. [b] This may include recognition of aspects where the organization is partially meeting a maturity level.				

附属書A (参考)

表 A.22 (続き)

細分箇条	成熟度レベル			結論	
	レベル	項目 a)	YES	結果／コメント b)
	5	利用可能な情報を収集し，分析し，レビューし，組織の状況，方針，戦略及び目標を更新し，理解する必要性を明確にし，組織のリーダーシップ活動の改善，学習及び革新の機会を特定するための体系的なアプローチを確立している．		

注 a) レベル3からレベル5までに記載している項目は，当該細分箇条に示した手引に基づいており，成熟度のレベルを示すことを意図している．
b) 結果／コメント欄は，組織が一つの成熟度レベルを部分的に満たしているという状況の認識を含んでもよい．

Table A.23 — Self-assessment of the detailed elements of 10.2

Subclause	Maturity level		Conclusion	
	Level	Item[a]	YES	Results/comment[b]
10.2 Performance indicators	1	Only basic performance indicators (e.g. financial criteria, on-time deliveries, number of customer complaints, legal warnings, fines) are used.		
		Data are not always reliable.		
	2	There is a limited set of performance indicators related to the organization's policies, strategy and objectives, and main processes.		
		Performance indicators are mostly based on the use of internal data.		
		Decisions are partially supported by measurable key performance indicators (KPIs).		
	3	Progress in achieving planned results against the policies, strategy and objectives in relevant processes and functions is identified and tracked by practical performance indicators.		
		The needs and expectations of customers and other interested parties are taken into account when selecting measurable KPIs.		
		Decisions are adequately supported by reliable, usable and measurable KPIs.		

附属書A（参考）

表A.23 — 10.2に対する自己評価

細分箇条	成熟度レベル		結論	
	レベル	項目 a)	YES	結果／コメント b)
10.2 パフォーマンス指標	1	基本的なパフォーマンス指標（例えば，財務基準，納期遵守，顧客苦情件数，法律上の警告，罰金）しか利用していない．		
		データが必ずしも信頼できるものではない．		
	2	組織の方針，戦略及び目標並びに主なプロセスに関連する，限られた一連のパフォーマンス指標がある．		
		パフォーマンス指標は，主に内部データに基づいている．		
		意思決定は，測定可能な主要パフォーマンス指標（KPI）によって部分的に支えられている．		
	3	関連するプロセス及び部門において，方針，戦略及び目標に照らし，計画した結果を達成できているかの進捗状況を，実践的なパフォーマンス指標によって特定し，追跡している．		
		測定可能なKPIの選定に際して，顧客及びその他の利害関係者のニーズ及び期待を考慮している．		
		意思決定は，信頼でき，使用に適しており，測定可能なKPIによって適切に支えている．		

Table A.23 *(continued)*

Subclause	Maturity level		Conclusion	
	Level	Item[a]	YES	Results/comment[b]
	4	Measurable KPIs are systematically selected to monitor progress in achieving planned results against the mission, vision, policies, strategy and objectives, at all levels and in all relevant processes and functions in the organization, to gather and provide the information necessary for performance evaluations and effective decision making.		
		Measurable KPIs provide information that is accurate, reliable and usable, in order to implement action plans when performance does not conform to objectives, or to improve and innovate process efficiency and effectiveness.		
	5	A process has been established to monitor progress in achieving planned results and make decisions using measurable KPIs.		
		Measurable KPIs contribute to good strategic and tactical decisions.		
		Information relating to risks and opportunities is considered when selecting measurable KPIs.		

[a] Items outlined in levels 3 to 5 are intended to be a progression of thought that is based on the guidance provided in the applicable subclause.
[b] This may include recognition of aspects where the organization is partially meeting a maturity level.

表 A.23（続き）

細分箇条	成熟度レベル		結論	
	レベル	項目 a)	YES	結果／コメント b)
	4	測定可能な KPI を体系的に選択して，組織の全ての階層及び全ての関係するプロセス及び部門において，使命，ビジョン，方針，戦略及び目標に照らし，計画した結果を達成できているかどうか進捗状況を監視し，パフォーマンス評価及び効果的な意思決定のために必要な情報を収集し，提供している．		
		測定可能な KPI が，パフォーマンスが目標に適合しない場合に実施計画を行うための，又はプロセスの効率及び有効性を改善し，刷新するための，正確で，信頼でき，使用に適した情報を提供している．		
	5	測定可能な KPI を使用して，計画した結果を達成できているかどうか進捗状況を監視し，意思決定を行うためのプロセスを確立している．		
		測定可能な KPI が優れた戦略的及び戦術的な決定に貢献している．		
		測定可能な KPI の選定に際して，リスク及び機会に関する情報を考慮している．		

注 a) レベル3からレベル5までに記載している項目は，当該細分箇条に示した手引に基づいており，成熟度のレベルを示すことを意図している．
 b) 結果／コメント欄は，組織が一つの成熟度レベルを部分的に満たしているという状況の認識を含んでもよい．

Table A.24 — Self-assessment of the detailed elements of 10.3

Subclause	Maturity level		Conclusion	
	Level	Item[a]	YES	Results/comment[b]
10.3 Performance analysis	1	The organization's performance is analysed in an informal or ad hoc manner.		
	2	There is limited analysis of the organization's performance.		
		Some basic statistical tools are used.		
	3	The organization's performance is analysed to identify issues and potential opportunities.		
		A systematic analysis process is supported by the wide use of statistical tools.		
	4	The organization's performance is analysed: — to identify insufficient resources; — to identify insufficient or ineffective competences, organizational knowledge and inappropriate behaviour; — to determine the new organizational knowledge needed; — to identify processes and activities showing outstanding performance that could be used as a model to improve other processes.		
		The effectiveness of the analysis process is enhanced by the sharing of the analysis results with interested parties.		

表A.24 − 10.3 に対する自己評価

細分箇条	成熟度レベル		結論	
	レベル	項目 a)	YES	結果／コメント b)
10.3 パフォーマンス分析	1	組織のパフォーマンスを，非公式又はその場限りの方法で分析している．		
	2	組織のパフォーマンスの分析は限られている．		
		幾つかの基本的な統計ツールを利用している．		
	3	課題及び潜在的な機会を特定するため，組織のパフォーマンスを分析している．		
		体系的な分析プロセスを，統計ツールの幅広い利用によって支えている．		
	4	組織のパフォーマンスを，次のような目的で分析している． — 不十分な資源の特定 — 不十分又は非効果的な力量及び組織の知識，並びに不適切な行動の特定 — 必要とする新しい組織の知識の明確化 — 不適切な行為の明確化 — 他のプロセスを改善するためのモデルとして使用することができる，傑出したパフォーマンスを示すプロセス及び活動の特定		
		分析プロセスの有効性を，利害関係者と分析結果を共有することによって高めている．		

Table A.24 *(continued)*

Subclause	Maturity level		Conclusion	
	Level	Item[a]	YES	Results/comment[b]
	5	The organization's performance is comprehensively analysed to identify potential strengths to be fostered with regard to the organization's leadership activities, as well as weakness in the organization's leadership roles and activities, including: — policy establishment and communication; — management of processes; — management of resources; — improvement, learning and innovation.		
		For the analysis, a clear framework to demonstrate the interrelations between its leadership roles, activities and their effects on the organization's performance is used.		

[a] Items outlined in levels 3 to 5 are intended to be a progression of thought that is based on the guidance provided in the applicable subclause.
[b] This may include recognition of aspects where the organization is partially meeting a maturity level.

附属書A（参考）

表A.24（続き）

細分箇条	成熟度レベル		結論	
	レベル	項目 a)	YES	結果／コメント b)
	5	組織のリーダーシップの役割及び活動における弱みだけでなく，リーダーシップ活動に関して伸ばす必要のありそうな潜在的な強みを特定するため，次の事項を含めて組織のパフォーマンスを総合的に分析している． ― 方針の策定及びコミュニケーション ― プロセスのマネジメント ― 資源のマネジメント ― 改善，学習及び革新		
		分析に当たって，組織のリーダーシップの役割，活動とそれらが組織のパフォーマンスに与える影響との相互関係を実証するための，明確な枠組みを使用している．		
注 a) レベル3からレベル5までに記載している項目は，当該細分箇条に示した手引に基づいており，成熟度のレベルを示すことを意図している．				
b) 結果／コメント欄は，組織が一つの成熟度レベルを部分的に満たしているという状況の認識を含んでもよい．				

Table A.25 — Self-assessment of the detailed elements of 10.4

Subclause	Maturity level		Conclusion	
	Level	Item[a]	YES	Results/comment[b]
10.4 Performance evaluation	1	The organization's performance is evaluated in an informal or ad hoc manner.		
	2	There is limited evaluation of the organization's performance.		
		Top management supports the identification and promulgation of best practices.		
		Some products from key competitors are evaluated and compared.		
	3	The results achieved on the organization's performance are evaluated against the applicable objectives.		
		The organization's performance is evaluated from the viewpoint of the needs and expectations of customers.		
		The organization's performance is evaluated using comparisons to established or agreed benchmarks.		
	4	Where the objectives have not been attained, the causes are investigated with appropriate review of the deployment of the organization's policies, strategy and objectives and the organization's managing of resources.		

附属書 A (参考)

表 A.25 － 10.4 に対する自己評価

細分箇条	成熟度レベル		結論	
	レベル	項目 a)	YES	結果／コメント b)
10.4 パフォーマンス評価	1	組織のパフォーマンスを，非公式又はその場限りの方法で評価している．		
	2	組織のパフォーマンスの評価は限られている．		
		トップマネジメントが，ベストプラクティスの特定及び普及を支持している．		
		主要な競合他社の幾つかの製品を分析し，比較している．		
	3	組織のパフォーマンスに関して達成した結果を，該当する目標に照らして評価している．		
		組織のパフォーマンスを，顧客のニーズ及び期待という視点から評価している．		
		組織のパフォーマンスを，確立した又は合意したベンチマークとの比較を用いながら評価している．		
	4	目標が達成されない場合には，その原因を調査し，組織の方針，戦略及び目標の展開，並びに組織の資源のマネジメントについて，適切なレビューを行っている．		

Table A.25 *(continued)*

Subclause	Maturity level		Conclusion	
	Level	Item[a]	YES	Results/ comment[b]
		The results of evaluation are understood comprehensively, and resolution of any identified gaps is prioritized based on their impacts on the organization's policies, strategy and objectives.		
		Improvement achieved on the organization's performance is evaluated from a long-term perspective.		
		The organization's performance is evaluated from the viewpoint of the needs and expectations of all interested parties.		
	5	Benchmarking is used systematically as a tool for identifying opportunities for improvement, learning and innovation.		
		The organization is frequently solicited by external entities to be a benchmark partner.		

[a] Items outlined in levels 3 to 5 are intended to be a progression of thought that is based on the guidance provided in the applicable subclause.

[b] This may include recognition of aspects where the organization is partially meeting a maturity level.

附属書A（参考）

表A.25（続き）

細分箇条	成熟度レベル		結論	
	レベル	項目 a)	YES	結果／コメント b)
		評価の結果を総合的に理解しており，組織の方針，戦略及び目標に対する影響に基づき，特定したギャップの解消を優先付けしている．		
		組織のパフォーマンスについて達成した改善を，長期的な展望から評価している．		
		組織のパフォーマンスを，全ての利害関係者のニーズ及び期待という視点から評価している．		
	5	ベンチマーキングを，改善，学習及び革新の機会を特定するためのツールとして体系的に利用している．		
		組織は，外部団体からベンチマークのパートナとなるように頻繁に求められている．		

注 a) レベル3からレベル5までに記載している項目は，当該細分箇条に示した手引に基づいており，成熟度のレベルを示すことを意図している．
  b) 結果／コメント欄は，組織が一つの成熟度レベルを部分的に満たしているという状況の認識を含んでもよい．

Table A.26 — Self-assessment of the detailed elements of 10.5

Subclause	Maturity level		Conclusion	
	Level	Item[a]	YES	Results/comment[b]
10.5 Internal audit	1	Internal audits are reactively performed in response to problems, customer complaints, etc.		
		Collected data are mostly used to resolve problems with products and services.		
	2	Internal audits for key processes are performed on a regular basis.		
		Collected data are used systematically to review the managing of processes.		
		Collected data are beginning to be used in a preventive way.		
	3	Internal audits are performed in a consistent manner, by competent personnel who are not involved in the activity being examined, in accordance with an audit plan.		
		Internal auditing identifies problems, nonconformities and risks, as well as monitoring progress in closing previously identified problems, nonconformities and risks.		
	4	Problems, nonconformities and risks identified are analysed comprehensively to determine weaknesses in the management system.		

附属書A（参考）

表A.26 − 10.5 に対する自己評価

細分箇条	成熟度レベル		結論	
	レベル	項目 a)	YES	結果／コメント b)
10.5 内部監査	1	内部監査を，問題，顧客からの苦情などに応じて受身的に実施しているだけである．		
		収集されたデータを，主に製品及びサービスに関する問題を解決するために利用している．		
	2	重要なプロセスに関する内部監査を，定期的に実施している．		
		収集されたデータを，プロセスのマネジメントについてのレビューを行うために体系的に利用している．		
		収集されたデータを，予防的にも使用し始めている．		
	3	内部監査を，監査計画に従って，評価の対象となっている活動に関与していない力量のある人々が，整合性のある方法で実施している．		
		以前に特定した問題，不適合及びリスクの終結に関する進捗状況を監視するだけでなく，内部監査によって，問題，不適合及びリスクを特定している．		
	4	特定した問題，不適合及びリスクを総合的に分析し，マネジメントシステムにおける弱みを明確にしている．		

Table A.26 *(continued)*

Subclause	Maturity level		Conclusion	
	Level	Item[a]	YES	Results/comment[b]
		Internal auditing focuses on the identification of good practices (which can be considered for use in other areas of the organization) as well as on improvement opportunities.		
	5	A process is established for the review of all internal audit reports to identify trends that can require organization-wide corrective actions or opportunities for improvement.		
		The organization involves other interested parties in its audits, in order to help identify additional opportunities for improvement.		

[a] Items outlined in levels 3 to 5 are intended to be a progression of thought that is based on the guidance provided in the applicable subclause.
[b] This may include recognition of aspects where the organization is partially meeting a maturity level.

表 A.26 (続き)

細分箇条	成熟度レベル		結論	
	レベル	項目 a)	YES	結果／コメント b)
		内部監査では,改善の機会だけでなく,(組織の他の分野への展開が考えられる)優れた実践の特定に焦点を合わせている.		
	5	組織全体にわたる是正処置を必要とするような傾向及び改善の機会を特定するために,全ての内部監査報告書をレビューするプロセスを確立している.		
		他の利害関係者を監査に参画させ,更なる改善の機会の特定に役立てている.		

注 a) レベル3からレベル5までに記載している項目は,当該細分箇条に示した手引に基づいており,成熟度のレベルを示すことを意図している.
b) 結果／コメント欄は,組織が一つの成熟度レベルを部分的に満たしているという状況の認識を含んでもよい.

Table A.27 — Self-assessment of the detailed elements of 10.6

Subclause	Maturity level		Conclusion	
	Level	Item[a]	YES	Results/comment[b]
10.6 Self-assessment	1	Self-assessment is not implemented.		
	2	Self-assessment is limited, informal or ad hoc.		
	3	Self-assessment is conducted in a consistent manner and the results are used to determine the organization's maturity and to improve its overall performance.		
	4	Self-assessment is used to determine the strengths and weaknesses of the organization, as well as its best practices, both at an overall level and at the level of individual processes.		
		Self-assessment assists the organization to prioritize, plan and implement improvements and/or innovations.		
	5	Self-assessment is performed by the organization at all levels.		
		The elements of a management system are understood comprehensively, based on the relations between the elements and their impacts on the organization's mission, vision, values and culture.		

附属書A (参考)

表A.27 − 10.6 に対する自己評価

細分箇条	成熟度レベル		結論	
	レベル	項目 a)	YES	結果／コメント b)
10.6 自己評価	1	自己評価を実施していない．		
	2	自己評価は，限られているか，非公式又はその場限りのものである．		
	3	自己評価を整合性のある方法で実施しており，その結果を組織の成熟度の判定及び組織の全体的なパフォーマンスの改善に利用している．		
	4	自己評価を，組織全体のレベル及び個々のプロセスレベルの両方における，組織の強み・弱み及びそのベストプラクティスを明確にするために利用している．		
		自己評価は，組織が改善及び／又は革新の優先順位を付け，計画し，実施することを手助けとなっている．		
	5	組織の全ての階層において自己評価を行っている．		
		マネジメントシステムの要素を，要素と，要素が組織の使命，ビジョン，価値観及び文化に与える影響との関係に基づいて総合的に理解している．		

Table A.27 *(continued)*

Subclause	Maturity level		Conclusion	
	Level	Item[a]	YES	Results/comment[b]
		The results of self-assessment are communicated to relevant people in the organization and used to share understanding about the organization and its future direction.		

[a] Items outlined in levels 3 to 5 are intended to be a progression of thought that is based on the guidance provided in the applicable subclause.
[b] This may include recognition of aspects where the organization is partially meeting a maturity level.

表 A.27（続き）

細分箇条	成熟度レベル		結論	
	レベル	項目 a)	YES	結果／コメント b)
		自己評価の結果を，組織内の関連する人々に伝達し，組織及びその今後の方向性についての理解を共有するために利用している．		

注 a) レベル3からレベル5までに記載している項目は，当該細分箇条に示した手引に基づいており，成熟度のレベルを示すことを意図している．
 b) 結果／コメント欄は，組織が一つの成熟度レベルを部分的に満たしているという状況の認識を含んでもよい．

Table A.28 — Self-assessment of the detailed elements of 10.7

Subclause	Maturity level		Conclusion	
	Level	Item[a]	YES	Results/comment[b]
10.7 Review	1	There is an ad hoc approach to reviews.		
		When a review is performed, it is often reactive.		
	2	Reviews are conducted to assess progress in the achievement of policies, strategy and objectives, and to assess the performance of the management system.		
		Relevant projects and improvement actions are assessed during reviews, in order to evaluate progress against their plans and objectives.		
	3	Systematic reviews of measurable KPIs and related objectives are undertaken at planned and periodic intervals, to enable trends to be determined, as well as to evaluate the organization's progress towards achieving its policies, strategy and objectives.		
		Where adverse trends are identified, they are acted upon.		
		Reviews enable evidence-based decision making.		

附属書 A（参考）

表 A.28 − 10.7 に対する自己評価

細分箇条	成熟度レベル			結論	
	レベル	項目 a)	YES	結果／コメント b)
10.7 レビュー	1	レビューの方法が，その場限りである．		
		多くの場合，問題対応形でレビューを行っている．		
	2	方針，戦略及び目標の達成における進捗状況を評価し，マネジメントシステムのパフォーマンスを評価するためのレビューを実施している．		
		レビューにおいて，関連するプロジェクト及び改善処置について，それらの計画及び目標に対する進捗状況を評価している．		
	3	傾向が明確になるよう，また，組織の方針，戦略及び目標の達成へ向けた進捗状況を評価するため，測定可能な KPI 及び該当する目標についての体系的なレビューを，あらかじめ定めた，定期的な間隔で実施している．		
		良くない傾向が特定された場合には，それらに対する処置をとっている．		
		レビューによって，証拠に基づく意思決定が可能になっている．		

Table A.28 *(continued)*

Subclause	Maturity level		Conclusion	
	Level	Item[a]	YES	Results/comment[b]
	4	The information, resulting from performance measurement, benchmarking, analysis and evaluations, internal audits and self-assessments, is comprehensively reviewed to identify opportunities for improvement, learning and innovation, as well as for identifying any need to adapt the organization's policies, strategy and objectives.		
		The outputs from the reviews are shared with interested parties, as a way of facilitating collaboration and learning.		
	5	Systematic reviews are used to identify opportunities for improvement, learning and innovation of the organization's leadership activities.		

[a] Items outlined in levels 3 to 5 are intended to be a progression of thought that is based on the guidance provided in the applicable subclause.
[b] This may include recognition of aspects where the organization is partially meeting a maturity level.

附属書A（参考）　　　　　　　　283

表 A.28（続き）

細分箇条	成熟度レベル		結論	
	レベル	項目 a)	YES	結果／コメント b)
	4	パフォーマンスの測定，ベンチマーキング，分析及び評価，内部監査並びに自己評価から得られる情報を総合的にレビューし，改善，学習及び革新の機会を特定するとともに，組織の方針，戦略及び目標を適応させるあらゆる必要性を特定している．		
		レビューから得られたアウトプットを，協働及び学習を促進させる方法として利害関係者と共有している．		
	5	体系的レビューを，組織のリーダーシップ活動の改善，学習及び革新の機会を特定するために使用している．		

注 a) レベル3からレベル5までに記載している項目は，当該細分箇条に示した手引に基づいており，成熟度のレベルを示すことを意図している．
b) 結果／コメント欄は，組織が一つの成熟度レベルを部分的に満たしているという状況の認識を含んでもよい．

Table A.29 — Self-assessment of the detailed elements of 11.1

Subclause	Maturity level		Conclusion	
	Level	Item[a]	YES	Results/comment[b]
11.1 Improvement, learning and innovation — General	1	Improvement activities are done in an informal or ad hoc manner.		
	2	Basic improvement processes, including corrections and corrective actions are in place, based on complaints from interested parties.		
	3	Improvement, learning and innovation efforts can be demonstrated in most products and some key processes.		
	4	Processes are implemented for the ongoing monitoring of external and internal issues that could lead to improvement, learning and innovation, which are aligned with strategic goals.		
	5	Improvement, learning and innovation are embedded as routine activities across the whole organization and are evident in relationships with interested parties.		

[a] Items outlined in levels 3 to 5 are intended to be a progression of thought that is based on the guidance provided in the applicable subclause.
[b] This may include recognition of aspects where the organization is partially meeting a maturity level.

表A.29 − 11.1 に対する自己評価

細分箇条	成熟度レベル		結論	
	レベル	項目 a)	YES	結果／コメント b)
11.1 改善，学習及び革新－一般	1	改善活動は，非公式又はその場限りのものである．		
	2	利害関係者からの苦情に基づいた，修正及び是正処置を含む基本的な改善プロセスを整備している．		
	3	大部分の製品及び幾つかの重要なプロセスにおいて，改善，学習及び革新の努力を実証している．		
	4	戦略的目標と一貫した改善，学習及び革新につながり得る外部及び内部の課題の継続的な監視のプロセスを実施している．		
	5	改善，学習及び革新を，組織全体にわたって日常活動として組み込んでおり，利害関係者との関係においてそれが明らかである．		

注 a) レベル3からレベル5までに記載している項目は，当該細分箇条に示した手引に基づいており，成熟度のレベルを示すことを意図している．
b) 結果／コメント欄は，組織が一つの成熟度レベルを部分的に満たしているという状況の認識を含んでもよい．

Table A.30 — Self-assessment of the detailed elements of 11.2

Subclause	Maturity level		Conclusion	
	Level	Item[a]	YES	Results/comment[b]
11.2 Improvement	1	Improvement activities are done in an informal or ad hoc manner.		
		Necessary resources to achieve improvement are provided.		
	2	Objectives for the improvement of products or services and processes are provided.		
		A structured approach is applied consistently.		
	3	The focus of improvement processes is aligned with the strategy and objectives, and top management is visibly involved in improvement activities.		
		Schemes are in place to empower teams and individuals to generate strategically relevant improvements.		
		Continual improvement processes include relevant interested parties.		
	4	Improvements and innovation result in learning and further improvements.		
	5	The focus of performance improvement is the sustained ability to learn, change and achieve long-term success.		

[a] Items outlined in levels 3 to 5 are intended to be a progression of thought that is based on the guidance provided in the applicable subclause.
[b] This may include recognition of aspects where the organization is partially meeting a maturity level.

附属書A（参考）

表A.30 — 11.2 に対する自己評価

細分箇条	成熟度レベル		結論	
	レベル	項目 a)	YES	結果／コメント b)
11.2 改善	1	改善活動は，非公式又はその場限りのものである．		
		改善を達成するために必要な資源は提供している．		
	2	製品又はサービスの改善のための目標及びプロセスを提供している．		
		構造化されたアプローチを整合性のある方法で適用している．		
	3	改善プロセスの焦点が戦略及び目標と一貫しており，トップマネジメントが目に見えるような形で改善活動に参画している．		
		チーム及び個人に戦略的に重要な改善を生み出す権限を委譲する制度を整備している．		
		継続的改善プロセスには，密接に関連する利害関係者が含まれている．		
	4	改善及び革新が学習及び一層の改善をもたらしている．		
	5	パフォーマンス改善の焦点が，長期的な成功を学習，変化及び達成する持続的な能力に当てられている．		

注 a) レベル3からレベル5までに記載している項目は，当該細分箇条に示した手引に基づいており，成熟度のレベルを示すことを意図している．
b) 結果／コメント欄は，組織が一つの成熟度レベルを部分的に満たしているという状況の認識を含んでもよい．

Table A.31 — Self-assessment of the detailed elements of 11.3

Subclause	Maturity level		Conclusion	
	Level	Item[a]	YES	Results/comment[b]
11.3 Learning	1	Some lessons are learned as a result of complaints.		
		Learning is on an individual basis, without the sharing of knowledge.		
	2	Learning is generated in a reactive way from the systematic analysis of problems and other information.		
		Processes exist for the sharing of information and knowledge, but still in a re-active manner.		
	3	Top management supports initiatives for learning and leads by example.		
		There are planned activities, events and forums for sharing information.		
		Processes are implemented to determine knowledge gaps and to provide the necessary resources for learning to occur.		
		Systems are in place for recognizing positive results from suggestions and lessons learned.		
	4	Learning is addressed in the strategy and policies.		
		Learning is recognized as a key issue.		
		Networking, connectivity and interactivity are stimulated by top management to share knowledge.		

表 A.31 − 11.3 に対する自己評価

細分箇条	成熟度レベル		結論	
	レベル	項目 a)	YES	結果／コメント b)
11.3 学習	1	苦情の結果として，幾つかの教訓を学習している．		
		知識を共有することはなく，学習が個人ベースのものとなっている．		
	2	学習を，問題及びその他の情報の，決めたとおりの分析から受身的に行っている．		
		情報及び知識を共有するプロセスが存在しているが，依然として受け身的である．		
	3	トップマネジメントが学習への取組みを支援し，手本となっている．		
		情報を共有するための計画的な活動，イベント及びフォーラムがある．		
		知識ギャップを明確にし，学習を行うために必要な資源を提供するためのプロセスを実施している．		
		提案及び教訓によって得られた良好な結果を表彰する制度を整備している．		
	4	戦略及び方針に学習を取り上げている．		
		学習を重要な課題として認識している．		
		知識を共有するため，トップマネジメントがネットワーク作り，人々のつながり及び相互作用を奨励している．		

Table A.31 *(continued)*

Subclause	Maturity level		Conclusion	
	Level	Item[a]	YES	Results/comment[b]
	5	The organization's learning ability integrates personal competence and the organization's overall competence.		
		Learning is fundamental to the improvement and innovation processes.		
		The organization's culture permits the taking of risks and learning from the mistakes.		
		There are external engagements for the purpose of learning.		

[a] Items outlined in levels 3 to 5 are intended to be a progression of thought that is based on the guidance provided in the applicable subclause.
[b] This may include recognition of aspects where the organization is partially meeting a maturity level.

表 A.31(続き)

細分箇条	成熟度レベル		結論	
	レベル	項目 a)	YES	結果／コメント b)
	5	組織の学習能力を通して,個人の力量と組織の全体的な力量とを統合している.		
		学習が改善及び革新プロセスのための基盤となっている.		
		リスクをとり,誤りから学ぶことを許容する組織文化がある.		
		学習のための外部との積極的参加が行われている.		

注 a) レベル3からレベル5までに記載している項目は,当該細分箇条に示した手引に基づいており,成熟度のレベルを示すことを意図している.
b) 結果／コメント欄は,組織が一つの成熟度レベルを部分的に満たしているという状況の認識を含んでもよい.

Table A.32 — Self-assessment of the detailed elements of 11.4

Subclause	Maturity level		Conclusion	
	Level	Item[a]	YES	Results/comment[b]
11.4 Innovation	1	There is limited innovation.		
		New products and services are introduced with no planning of the innovation process.		
	2	Innovation activities are based on data relating to the needs and expectations of interested parties.		
	3	The innovation processes for new products and services are able to identify changes in external and internal issues, in order to plan innovations.		
		Risks associated with planned innovations are considered.		
		The organization supports the innovation initiatives with the resources needed.		
	4	Innovations are prioritized, with balanced consideration of urgency, availability of resources and the organization's strategy.		
		External providers and partners are involved in innovation processes.		
		The effectiveness and efficiency of innovation processes are assessed regularly as a part of the learning process.		

表 A.32 − 11.4 に対する自己評価

細分箇条	成熟度レベル		結論	
	レベル	項目 a)	YES	結果／コメント b)
11.4 革新	1	革新がほとんど行われていない．		
		新規製品及びサービスを，革新プロセスの計画なしに導入している．		
	2	革新活動は，利害関係者のニーズ及び期待に関するデータに基づいて行われている．		
	3	新規製品及びサービスのための革新プロセスは，革新を計画するために必要な外部及び内部の課題の変化を特定することができるものである．		
		計画された革新に伴うリスクを考慮している．		
		組織は，必要となる資源を用いて革新への取組みを支援している．		
	4	革新の優先順位付けを，革新の緊急性，資源の利用可能性，及び組織の戦略のバランスのとれた考慮とともに行っている．		
		外部提供者及びパートナが革新プロセスに参画している．		
		革新プロセスの有効性及び効率の定期的な評価を，学習プロセスの一環として行っている．		

Table A.32 *(continued)*

Subclause	Maturity level		Conclusion	
	Level	Item[a]	YES	Results/comment[b]
		Innovation is used to improve the way the organization operates.		
	5	Innovation activities anticipate possible changes in the context of the organization.		
		Preventive plans are developed to avoid or minimize the identified risks that accompany the innovation activities.		
		Innovation is applied at all levels, through changes in technology, processes, organization, the management system and the organization's business model.		

[a] Items outlined in levels 3 to 5 are intended to be a progression of thought that is based on the guidance provided in the applicable subclause.
[b] This may include recognition of aspects where the organization is partially meeting a maturity level.

表 A.32 （続き）

細分箇条	成熟度レベル		結論	
	レベル	項目 a)	YES	結果／コメント b)
		革新を組織の運営方法の改善に利用している．		
	5	革新活動が，組織の状況における起こり得る変化を予想したものとなっている．		
		革新活動に伴い特定したリスクを回避又は最小限に抑えるための予防計画を策定している．		
		技術，プロセス，組織，マネジメントシステム及び組織のビジネスモデルにおける変化を通じて，革新を全階層で適用している．		

注 a) レベル3からレベル5までに記載している項目は，当該細分箇条に示した手引に基づいており，成熟度のレベルを示すことを意図している．
 b) 結果／コメント欄は，組織が一つの成熟度レベルを部分的に満たしているという状況の認識を含んでもよい．

Bibliography

[1] ISO 9001, *Quality management systems — Requirements*

[2] ISO/TS 9002, *Quality management systems — Guidelines for the application of ISO 9001:2015*

[3] ISO 10001, *Quality management — Customer satisfaction — Guidelines for codes of conduct for organizations*

[4] ISO 10002, *Quality management — Customer satisfaction — Guidelines for complaints handling in organizations*

[5] ISO 10003, *Quality management — Custom-*

参考文献

[1] **JIS Q 9001** 品質マネジメントシステム―要求事項

　　注記 対応国際規格：**ISO 9001**, Quality management systems―Requirements

[2] **JIS Q 9002** 品質マネジメントシステム―JIS Q 9001 の適用に関する指針

　　注記 対応国際規格：**ISO/TS 9002**, Quality management systems―Guidelines for the application of ISO 9001:2015

[3] **JIS Q 10001** 品質マネジメント―顧客満足―組織における行動規範のための指針

　　注記 対応国際規格：**ISO 10001**, Quality management―Customer satisfaction―Guidelines for codes of conduct for organizations

[4] **JIS Q 10002** 品質マネジメント―顧客満足―組織における苦情対応のための指針

　　注記 対応国際規格：**ISO 10002**, Quality management ― Customer satisfaction ― Guidelines for complaints handling in organizations

[5] **JIS Q 10003** 品質マネジメント―顧客満足

er satisfaction — *Guidelines for dispute resolution external to organizations*

[6] ISO 10004, *Quality management — Customer satisfaction — Guidelines for monitoring and measuring*
[7] ISO 10005, *Quality management — Guidelines for quality plans*
[8] ISO 10006, *Quality management — Guidelines for quality management in projects*
[9] ISO 10007, *Quality management — Guidelines for configuration management*
[10] ISO 10008, *Quality management — Customer satisfaction — Guidelines for business-to-consumer electronic commerce transactions*
[11] ISO 10012, *Measurement management systems — Requirements for measurement processes and measuring equipment*

―組織の外部における紛争解決のための指針

 注記 対応国際規格：**ISO 10003**, Quality management―Customer satisfaction―Guidelines for dispute resolution external to organizations

[6] **ISO 10004**, Quality management―Customer satisfaction―Guidelines for monitoring and measuring

[7] **ISO 10005**, Quality management systems―Guidelines for quality plans

[8] **ISO 10006**, Quality management―Guidelines for quality management in projects

[9] **ISO 10007**, Quality management―Guidelines for configuration management

[10] **ISO 10008**, Quality management―Customer satisfaction―Guidelines for business-to-consumer electronic commerce transactions

[11] **JIS Q 10012** 計測マネジメントシステム―測定プロセス及び測定機器に関する要求事項

 注記 対応国際規格：**ISO 10012**, Measurement management systems―Requirements for measurement processes and measuring equipment

[12] ISO/TR 10013, *Guidelines for quality management system documentation*

[13] ISO 10014, *Quality management — Guidelines for realizing financial and economic benefits*

[14] ISO 10015, *Quality management — Guidelines for training*

[15] ISO 10018, *Quality management — Guidelines on people involvement and competence*

[16] ISO 10019, *Guidelines for the selection of quality management system consultants and use of their services*

[17] ISO 14001, *Environmental management systems — Requirements with guidance for use*

[18] ISO 14040, *Environmental management —*

- [12] **ISO/TR 10013**, Guidelines for quality management system documentation
- [13] **ISO 10014**, Quality management—Guidelines for realizing financial and economic benefits
- [14] **ISO 10015**, Quality management—Guidelines for training
- [15] **ISO 10018**, Quality management—Guidelines on people involvement and competence
- [16] **JIS Q 10019** 品質マネジメントシステムコンサルタントの選定及びそのサービスの利用のための指針

    注記 対応国際規格：**ISO 10019**, Guidelines for the selection of quality management system consultants and use of their services
- [17] **JIS Q 14001** 環境マネジメントシステム―要求事項及び利用の手引

    注記 対応国際規格：**ISO 14001**, Environmental management systems—Requirements with guidance for use
- [18] **JIS Q 14040** 環境マネジメント―ライフサ

Life cycle assessment — Principles and framework

[19] ISO 14044, *Environmental management — Life cycle assessment — Requirements and guidelines*

[20] ISO/TR 14047, *Environmental management — Life cycle assessment — Illustrative examples on how to apply ISO 14044 to impact assessment situations*

[21] ISO/TS 14048, *Environmental management — Life cycle assessment — Data documentation format*

[22] ISO/TR 14049, *Environmental management — Life cycle assessment — Illustrative examples on how to apply ISO 14044 to goal and scope definition and inventory analysis*

[23] ISO/TR 14062, *Environmental management*

イクルアセスメント—原則及び枠組み

  注記 対応国際規格:**ISO 14040**, Environmental management—Life cycle assessment—Principles and framework

[19] **JIS Q 14044** 環境マネジメント—ライフサイクルアセスメント—要求事項及び指針

  注記 対応国際規格:**ISO 14044**, Environmental management—Life cycle assessment—Requirements and guidelines

[20] **ISO/TR 14047**, Environmental management—Life cycle assessment—Illustrative examples on how to apply ISO 14044 to impact assessment situations

[21] **ISO/TS 14048**, Environmental management—Life cycle assessment—Data documentation format

[22] **ISO/TR 14049**, Environmental management—Life cycle assessment—Illustrative examples on how to apply ISO 14044 to goal and scope definition and inventory analysis

[23] **ISO/TR 14062**, Environmental manage-

— *Integrating environmental aspects into product design and development*

[24] ISO 19011, *Guidelines for auditing management systems*

[25] ISO 19600, *Compliance management systems — Guidelines*

[26] ISO 22000, *Food safety management systems — Requirements for any organization in the food chain*

[27] ISO 22301, *Societal security — Business continuity management systems — Requirements*

[28] ISO 22316, *Security and resilience — Organizational resilience — Principles and attributes*

[29] ISO 26000, *Guidance on social responsibility*

[30] ISO 31000, *Risk management — Guidelines*

ment—Integrating environmental aspects into product design and development

[24] **JIS Q 19011** マネジメントシステム監査のための指針

  注記 対応国際規格：**ISO 19011**, Guidelines for auditing management systems

[25] **ISO 19600**, Compliance management systems—Guidelines

[26] **ISO 22000**, Food safety management systems—Requirements for any organization in the food chain

[27] **JIS Q 22301** 社会セキュリティ―事業継続マネジメントシステム―要求事項

  注記 対応国際規格：**ISO 22301**, Societal security—Business continuity management systems—Requirements

[28] **ISO 22316**, Security and resilience—Organizational resilience—Principles and attributes

[29] **JIS Z 26000** 社会的責任に関する手引

  注記 対応国際規格：**ISO 26000**, Guidance on social responsibility

[30] **JIS Q 31000** リスクマネジメント―原則及

[31] ISO 37001, *Anti-bribery management systems — Requirements with guidance for use*

[32] ISO 39001, *Road traffic safety (RTS) management systems — Requirements with guidance for use*

[33] ISO 45001, *Occupational health and safety management systems — Requirements with guidance for use*

[34] ISO 50001, *Energy management systems — Requirements with guidance for use*

[35] ISO/IEC 27000, *Information technology — Security techniques — Information security management systems — Overview and vocabulary*

[36] ISO/IEC 27001, *Information technology —*

び指針

> **注記** 対応国際規格：**ISO 31000**, Risk management—Guidelines

[31] **ISO 37001**, Anti-bribery management systems—Requirements with guidance for use

[32] **ISO 39001**, Road traffic safety (RTS) management systems—Requirements with guidance for use

[33] **JIS Q 45001** 労働安全衛生マネジメントシステム—要求事項及び利用の手引

> **注記** 対応国際規格：**ISO 45001**, Occupational health and safety management systems—Requirements with guidance for use

[34] **JIS Q 50001** エネルギーマネジメントシステム—要求事項及び利用の手引

> **注記** 対応国際規格：**ISO 50001**, Energy management systems—Requirements with guidance for use

[35] **ISO/IEC 27000**, Information technology—Security techniques—Information security management systems—Overview and vocabulary

[36] **JIS Q 27001** 情報技術—セキュリティ技術

Security techniques — Information security management systems — Requirements

[37] ISO/IEC 27002, *Information technology — Security techniques — Code of practice for information security controls*

[38] ISO/IEC IEEE 24748-5, *Systems and software engineering — Life cycle management — Part 5: Software development planning*

[39] IEC 60300-1, *Dependability management — Part 1: Dependability management systems*

[40] IEC 61160, *Design review*

[41] ISO Handbook. *ISO 9001:2015 for Small Enterprises — What to do?* 2016. Available at: **https://www.iso.org/publication/PUB100406.html**

—情報セキュリティマネジメントシステム—要求事項

  注記 対応国際規格：**ISO/IEC 27001**, Information technology—Security techniques—Information security management systems—Requirements

[37] **JIS Q 27002**　情報技術—セキュリティ技術—情報セキュリティ管理策の実践のための規範

  注記 対応国際規格：**ISO/IEC 27002**, Information technology—Security techniques—Code of practice for information security controls

[38] **ISO/IEC IEEE 24748-5**, Systems and software engineering—Life cycle management—Part 5: Software development planning

[39] **IEC 60300-1**, Dependability management—Part 1: Guidance for management and application

[40] **IEC 61160**, Design review

[41] ISO Handbook. ISO 9001:2015 for Small Enterprises—What to do? 2016 https://www.iso.org/publication/PUB100406.html

[42] ISO. *Guidance on the Concept and Use of the Process Approach for management systems.* ISO/TC 176/SC 2/N 544R3, 2008. Available at: **https://www.iso.org/files/live/sites/isoorg/files/archive/pdf/en/04_concept_and_use_of_the_process_approach_for_management_systems.pdf**

[43] ISO information and guidance on ISO 9001 and ISO 9004. Available at: **https://committee.iso.org/tc176sc2**

[44] ISO 9001 *Auditing Practices Group. Various papers.* Available at **https://committee.iso.org/sites/tc176sc2/home/page/iso-9001-auditing-practices-grou.html**

参考文献

- [42] ISO. Guidance on the Concept and Use of the Process Approach for management systems. **ISO/TC 176**/SC 2/N 544R3, 2008
  https://www.iso.org/files/live/sites/isoorg/files/archive/pdf/en/04_concept_and_use_of_the_process_approach_for_management_systems.pdf
- [43] ISO information and guidance on ISO 9001 and ISO 9004
  https://committee.iso.org/tc176sc2
- [44] ISO 9001, Auditing Practices Group. Various papers
  https://committee.iso.org/sites/tc176sc2/home/page/iso-9001-auditing-practices-grou.html
- [45] **JIS Q 9023** マネジメントシステムのパフォーマンス改善―方針管理の指針
- [46] **JIS Q 9024** マネジメントシステムのパフォーマンス改善―継続的改善の手順及び技法の指針
- [47] **JIS Q 9026** マネジメントシステムのパフォーマンス改善―日常管理の指針
- [48] **JIS Q 9027** マネジメントシステムのパフォーマンス改善―プロセス保証の指針

対訳 ISO 9004:2018（JIS Q 9004:2018）
品質マネジメント―組織の品質―持続的成功を達成するための指針［ポケット版］

定価：本体 5,500 円(税別)

2019 年 3 月 1 日	第 1 版第 1 刷発行

編　　者　一般財団法人 日本規格協会

発 行 者　揖斐　敏夫

発 行 所　一般財団法人 日本規格協会

〒 108-0073　東京都港区三田 3 丁目 13-12 三田 MT ビル
　　　　　　http://www.jsa.or.jp/
　　　　　　振替　00160-2-195146

印 刷 所　株式会社ディグ

© Japanese Standards Association, et al., 2019　　Printed in Japan
ISBN978-4-542-40283-6

- 当会発行図書，海外規格のお求めは，下記をご利用ください．
  販売サービスチーム：(03)4231-8550
  書店販売：(03)4231-8553　注文 FAX：(03)4231-8665
  JSA Webdesk：https://webdesk.jsa.or.jp/